# Intelligent Systems, Control and Automation: Science and Engineering

Volume 104

**Series Editor**

Kimon P. Valavanis, Department of Electrical and Computer Engineering, University of Denver, Denver, CO, USA

**Advisory Editors**

P. Antsaklis, University of Notre Dame, Notre Dame, IN, USA

P. Borne, Ecole Centrale de Lille, France

R. Carelli, Universidad Nacional de San Juan, Argentina

T. Fukuda, Nagoya University, Japan

N. R. Gans, The University of Texas at Dallas, Richardson, TX, USA

F. Harashima, University of Tokyo, Japan

P. Martinet, Ecole Centrale de Nantes, France

S. Monaco, University La Sapienza, Rome, Italy

R. R. Negenborn, Delft University of Technology, Delft, The Netherlands

António Pascoal, Institute for Systems and Robotics, Lisbon, Portugal

G. Schmidt, Technical University of Munich, Germany

T. M. Sobh, University of Bridgeport, Bridgeport, CT, USA

Intelligent Systems, Control and Automation: Science and Engineering book series publishes books on scientific, engineering, and technological developments in this interesting field that borders on so many disciplines and has so many practical applications: human-like biomechanics, industrial robotics, mobile robotics, service and social robotics, humanoid robotics, mechatronics, intelligent control, industrial process control, power systems control, industrial and office automation, unmanned aviation systems, teleoperation systems, energy systems, transportation systems, driverless cars, human-robot interaction, computer and control engineering, but also computational intelligence, neural networks, fuzzy systems, genetic algorithms, neurofuzzy systems and control, nonlinear dynamics and control, and of course adaptive, complex and self-organizing systems. This wide range of topics, approaches, perspectives and applications is reflected in a large readership of researchers and practitioners in various fields, as well as graduate students who want to learn more on a given subject.

The series has received an enthusiastic acceptance by the scientific and engineering community, and is continuously receiving an increasing number of high-quality proposals from both academia and industry. The current Series Editor is Kimon Valavanis, University of Denver, Colorado, USA. He is assisted by an Editorial Advisory Board who help to select the most interesting and cutting edge manuscripts for the series:

Panos Antsaklis, University of Notre Dame, USA
Stjepan Bogdan, University of Zagreb, Croatia
Alexandre Brandao, UFV, Brazil
Giorgio Guglieri, Politecnico di Torino, Italy
Kostas Kyriakopoulos, National Technical University of Athens, Greece
Rogelio Lozano, University of Technology of Compiegne, France
Anibal Ollero, University of Seville, Spain
Hai-Long Pei, South China University of Technology, China
Tarek Sobh, University of Bridgeport, USA

Springer and Professor Valavanis welcome book ideas from authors. Potential authors who wish to submit a book proposal should contact Thomas Ditzinger (thomas.ditzinger@springer.com)

Indexed by SCOPUS, zbMATH, SCImago.

Rihard Karba · Juš Kocijan · Tadej Bajd ·
Mojca Žagar Karer · Gorazd Karer

# Terminological Dictionary of Automatic Control, Systems and Robotics

Rihard Karba
Faculty of Electrical Engineering
University of Ljubljana
Ljubljana, Slovenia

Tadej Bajd
Faculty of Electrical Engineering
University of Ljubljana
Ljubljana, Slovenia

Gorazd Karer
Faculty of Electrical Engineering
University of Ljubljana
Ljubljana, Slovenia

Juš Kocijan
Department of Systems and Control
Jožef Stefan Institute
Ljubljana, Slovenia

Centre for Information Technologies
and Applied Mathematics
University of Nova Gorica
Nova Gorica, Slovenia

Mojca Žagar Karer
Fran Ramovš Institute of the Slovenian
Language
Research Centre of the Slovenian Academy
of Sciences and Arts
Ljubljana, Slovenia

ISSN 2213-8986          ISSN 2213-8994  (electronic)
Intelligent Systems, Control and Automation: Science and Engineering
ISBN 978-3-031-35754-1          ISBN 978-3-031-35755-8  (eBook)
https://doi.org/10.1007/978-3-031-35755-8

Terminološki slovar avtomatike/Terminological Dictionary of Automatic Control, Systems and Robotics,
Copyright © ZRC SAZU/Research Centre of the Slovenian Academy of Sciences and Arts, 2014, under
licence to Springer International Publishing AG. All rights reserved.

© ZRC SAZU/Research Centre of the Slovenian Academy of Sciences and Arts 2023, corrected
publication 2024

This work is subject to copyright. All rights are reserved by the Publisher, whether the whole or part of the
material is concerned, specifically the rights of reprinting, reuse of illustrations, recitation, broadcasting,
reproduction on microfilms or in any other physical way, and transmission or information storage and
retrieval, electronic adaptation, computer software, or by similar or dissimilar methodology now known
or hereafter developed.
The use of general descriptive names, registered names, trademarks, service marks, etc. in this publication
does not imply, even in the absence of a specific statement, that such names are exempt from the relevant
protective laws and regulations and therefore free for general use.
The publisher, the authors, and the editors are safe to assume that the advice and information in this book
are believed to be true and accurate at the date of publication. Neither the publisher nor the authors or
the editors give a warranty, expressed or implied, with respect to the material contained herein or for any
errors or omissions that may have been made. The publisher remains neutral with regard to jurisdictional
claims in published maps and institutional affiliations.

This Springer imprint is published by the registered company Springer Nature Switzerland AG
The registered company address is: Gewerbestrasse 11, 6330 Cham, Switzerland

Paper in this product is recyclable.

The original version of these chapters has been revised: Abstracts in the online version of Chapters 2 A to 13 L have been updated. A corrections to these chapters can be found at https://doi.org/10.1007/978-3-031-35755-8 _28

# Acknowledgements

The authors want to express their sincere gratitude to all the colleagues covering different areas of the vocabulary scope, who, with their expertise, contributed to the dictionary entries and definitions. Also, the initiative of the Automatic Control Society of Slovenia and the collaboration of the Terminological Section of Fran Ramovš Institute of the Slovenian Language at the Research Centre of the Slovenian Academy of Sciences and Arts, as well as the support of the Laboratory of Control Systems and Cybernetics, Faculty of Electrical Engineering, University of Ljubljana, and Department of Systems and Control, Jožef Stefan Institute, are greatly appreciated.

# Contents

**Part I   Introduction**

1   **Automatic Control, Systems and Robotics** ........................ 3

**Part II   Terminological Dictionary of Automatic Control, Systems and Robotics**

2   **A** ....................................................... 11

3   **B** ....................................................... 23

4   **C** ....................................................... 31

5   **D** ....................................................... 49

6   **E** ....................................................... 67

7   **F** ....................................................... 75

8   **G** ....................................................... 85

9   **H** ....................................................... 91

10   **I** ....................................................... 99

11   **J** ....................................................... 111

12   **K** ....................................................... 113

13   **L** ....................................................... 115

14   **M** ....................................................... 123

15   **N** ....................................................... 137

16   **O** ....................................................... 143

17   **P** ....................................................... 151

18   **Q** ....................................................... 171

| 19 | R | 173 |
|----|---|-----|
| 20 | S | 193 |
| 21 | T | 213 |
| 22 | U | 225 |
| 23 | V | 231 |
| 24 | W | 239 |
| 25 | Y | 241 |
| 26 | Z | 243 |

**Part III   References**

27   **References and Terminological References** .......................   247

**Correction to: Terminological Dictionary of Automatic Control, Systems and Robotics** .............................................   C1

# List of Figures

| | | |
|---|---|---|
| Fig. 1 | Dead time (2) | 51 |
| Fig. 2 | Delay time (1) | 52 |
| Fig. 3 | Reaction curve | 176 |
| Fig. 4 | Reaction rate | 176 |
| Fig. 5 | Rise time (1) for overdamped second-order systems | 183 |
| Fig. 6 | Rise time (1) for underdamped second-order systems | 183 |
| Fig. 7 | Rise time (2) | 184 |

# Part I
# Introduction

# Chapter 1
# Automatic Control, Systems and Robotics

Automatic control studies methods of analysis and synthesis of automatic control systems. It deals with using automatic control devices and information technologies to improve the efficiency of production, services, and plant operations. The dictionary comprises terms from the automatic control field, including mathematical modelling, the simulation of dynamical systems, automation technology with relevant elements, and robotics. The field is also interconnected with signal processing, information, and production technologies.

Because of its systemic approach to problem-solving, automatic control is now a highly interdisciplinary science that is indispensable in many technical and non-technical fields, such as smart buildings and devices, advanced manufacturing systems, autonomous vehicles, aerospace engineering, in medicine, as well as in the study of various processes in biological, economic, and sociological systems. The introduction of automatic control reduces the consumption of raw materials and energy, enables better utilisation of production capacity, greater production flexibility, higher product quality, humanisation of work, increased safety of people and machines, and reduced environmental pollution. Furthermore, it is often a prerequisite for adequately functioning systems that would otherwise be useless or dangerous.

Knowledge of automatic control provides a better understanding of the often complex mechanisms of systems in various subject fields. It offers opportunities to influence behaviour effectively and thus improve many aspects of the systems in question. It is essential for experts in the field of automatic control, who must communicate clearly and unambiguously, to use consistent terminology. Therefore, we decided to compile an English terminological dictionary based on the Slovenian terminological dictionary "Terminološki slovar avtomatike" (ZRC SAZU, 2014 (1st ed.); 2018 (2nd ed.)). We have added new terms and significantly expanded the definitions, resulting in a new dictionary that we hope will be useful for various types of users, especially experts in the field of automatic control. The dictionary contains 2433 dictionary entries.

© ZRC SAZU/Research Centre of the Slovenian Academy of Sciences and Arts 2023
R. Karba et al., *Terminological Dictionary of Automatic Control, Systems and Robotics*,
Intelligent Systems, Control and Automation: Science and Engineering 104,
https://doi.org/10.1007/978-3-031-35755-8_1

We believe that a well-defined terminology ensures noiseless communication between professionals and thus gives a considerable boost to the development of the field. Feel free to address any remarks or suggestions regarding the dictionary to the corresponding author Gorazd Karer (gorazd.karer@fe.uni-lj.si).

## 1.1 About the Dictionary

When compiling a terminological dictionary, collaboration between a group of experts, who have extensive knowledge about the concepts, and a terminologist, who has experience in compiling terminological dictionaries, is essential. Four experts in automatic control and robotics, and one terminologist, were involved in this project. Each expert provided definitions of concepts and possible synonyms. The proposed definitions were discussed regularly with the other experts and the terminologist.

Since the terms occur mainly in technical texts, the relevant technical literature, i.e., scientific articles and monographs, university textbooks, technical articles and monographs, were consulted and systematically reviewed to compile the dictionary entries. The dictionary definitions were also checked in secondary sources, i.e. related terminological dictionaries and lexicons. All other available resources were also considered. An index of the essential literature used is provided at the end of the dictionary.

The dictionary is ordered alphabetically. It reflects the current state of terminology in this subject field and does not include terms, which are no longer used in practice. It also does not explain the historical development of concepts or the etymology of terms. It does not include the established terms from related subject fields, which are used in automatic control but their definition is the same as in the primary discipline (e.g., the mathematical term *Laplace transform*). The dictionary's authors assume that subject-field experts, who are the intended users of this dictionary, are familiar with these terms. However, due to the interdisciplinary nature of automatic control, some terms referring to related disciplines (e.g., *lean manufacturing*) are included.

## 1.2 Users of the Dictionary

The dictionary is primarily, but not exclusively, intended for experts and students working in control engineering and dynamic systems in technical and non-technical areas. Therefore, a basic knowledge of this field is required to use the dictionary. For students, the dictionary provides a sound and direct contact with the technical language that aspiring engineers must master to communicate effectively within their subject field. It is a helpful tool when writing final theses and reports by providing relevant information with definitions and information about the preferred terms.

## 1.3 Methodology

The Terminological Dictionary of Automatic Control, Systems and Robotics is based on the conceptual approach, which means that terms are treated as designations for concepts that make up the conceptual system of a discipline. For this reason, terminology work must consider the relations between concepts (terms). In the dictionary, these relations may be expressed explicitly, e.g., as a cross-reference indicating a synonymous pair (i.e., two terms for the same concept), or implicitly, e.g., as a hierarchical relation of superordinate and subordinate in a definition.

Terminology work typically involves grouping related terms together. For instance, all types of different level sensors (*capacitive level sensor, conductivity level sensor, ultrasonic level sensor, optical level sensor*, etc.) are discussed and defined together. This approach reduces the risk of conceptual inconsistencies and leads to more reliable definitions.

## 1.4 Dictionary Structure

The terms are part of a terminological system that reflects the conceptual system of the subject field. In the dictionary, each term introduces its own dictionary entry. A dictionary entry consists of a term, a definition, and (optionally) a synonym. The definition may consist of one sentence or more sentences, usually three or four. Examples:

**series compensation** Compensation, in which the compensator is connected to the input or to the output of the controlled system in the direct path of the control loop.

**solenoid** An actuator, which uses an electromagnet to generate linear displacement. A current signal in the coil causes the ferromagnetic core to be pushed or pulled towards the centre of the coil. When the excitation stops, the corresponding spring returns the ferromagnetic core to the starting position. It is commonly used in, e.g., valve, robot, actuator system, air conditioning system.

Sometimes the same term designates different concepts. In this case, the definitions are numbered. Example:

**signal** **1**. A physical quantity that conveys information, e.g., voltage, current, electromagnetic wave. It is time-dependent and position-dependent. **2**. An element of a block diagram or of a signal-flow graph, which shows a path of the connection between blocks with a line and its direction with an arrow.

Ideally, the definition should place the concept into the concept system. For this reason, related concepts have been defined together so that the definitions can be precise and consistent. For example, when defining different types of *control (adaptive control, multivariable control, hybrid control, automatic control, basic control,*

6                                       1 Automatic Control, Systems and Robotics

*fuzzy control*, etc.), all related concepts were considered. The definition often starts with a superordinate term. Example:

**valve** A device, which controls the flow of fluid or energy by directly affecting the controlled object. It can be either of electrical type, e.g., transistor, thyristor, rheostat, or process type, e.g., control valve, on-off valve, safety valve.

> **control valve** A valve, which changes the flow rate of the fluid according to the corresponding control signal, e.g., globe valve, gate valve, ball valve. It consists of a valve actuator, a valve positioner and a valve body enabling continuous setting of all states from completely opened to completely closed. It may provide the additional possibility of manual intervention. It is the most common final control element in process industry.
>
>> **direct acting control valve** A control valve, which tends to be opened when the flow increases. It requires a special construction of the valve actuator, as well as of the valve disc and the valve seat.
>>
>> **reverse acting control valve** A control valve, which tends to be closed when the flow rate increases. It requires a special construction of the valve actuator, as well as of the valve disc and the valve seat.

The terms from the field of automatic control, systems and robotics, which are part of a definition of another term, are defined in a separate dictionary entry. Example:

**resonant peak** The highest value of the amplitude response of a dynamic system in the Bode plot. It occurs at the resonant frequency as the consequence of a conjugate complex pair of poles in the open-loop transfer function.

**amplitude response** The part of frequency response, which is the ratio of the output-signal amplitude at a particular frequency and the input-signal amplitude at the same frequency. It is often given in decibels as a part of the Bode plot. **S**: magnitude response

**dynamic system** A system, in which time-dependent qualitative or quantitative changes are taking place. The output of such a system depends not only on the current input value but also on the previous input values. **S**: dynamical system

**Bode plot** A graphical representation of the frequency response, which is used for analysis and control-system design. It consists of the Bode magnitude plot and the Bode phase plot that are shown on separate graphs, both with the same logarithmic frequency scale along the $x$-axis. **S**: Bode diagram

**resonant frequency** The frequency of a harmonic input signal, at which the largest ratio between steady-state amplitudes of output signals and input signals is reached.

**pole 1**. A value of complex variable $s$ or complex variable $z$, which results in a zero-valued denominator of the continuous or discrete transfer function and therefore makes the transfer function singular. Its position in the complex plane affects the stability and behaviour of the system. **2**. An eigenvalue of the system matrix. Its position in the complex plane affects the stability and behaviour of the system. **3**. A root of the characteristic equation of a linear closed-loop system. Its position in the complex plane affects the stability and behaviour of the system.

**open-loop transfer function** A transfer function, which is the product of all transfer functions along the forward path and along the feedback path, connected in series.

A headword is always a term in its full form, e.g., *artificial neural network*. Short forms of terms, e.g., *neural network*, are not listed. However, abbreviations, e.g., *ANN*, are treated as synonyms (see the following section).

## 1.5 Cross-References

Some terms have one or more synonyms, i.e., terms that designate the same concept. In this case, each synonym is also listed in a separate dictionary entry.

When different terms are used for the same concept, this can lead to misunderstandings that prevent accurate and effective communication among experts. Since the dictionary attempts to harmonise terminology in the field of automatic control, systems and robotics, there is a distinction between the preferred terms and the non-preferred terms. The preferred terms are thus represented by full dictionary entries (with definitions). In contrast, non-preferred terms are represented by dictionary entries containing the term and an arrow ($\rightarrow$) referencing the preferred term instead of a definition. Usually, the most frequently used term is preferred, but other terminological principles have been considered as well (e.g., linguistic adequacy, preference for shorter terms). Sometimes a terminological agreement is required within a larger group of experts. The dictionary entries with preferred terms always include information about synonyms (they are listed after the character S). However, it should be noted that non-preferred terms are not considered incorrect. Example:

**optoisolator** $\rightarrow$ optocoupler

**photocoupler** $\rightarrow$ optocoupler

**optocoupler** A single-package semiconductor device, which transfers a signal between two galvanically isolated circuits by using light, with, e.g., light-emitting diode, infrared emitting diode, laser diode, phototransistor, photodiode. It can transfer digital signals as well as analogue signals in, e.g., control, monitoring, communication. **S**: optoisolator, photocoupler

Since abbreviations are very common in automatic control, their use is not discouraged. In the dictionary, an abbreviation is followed by a double-sided arrow ($\leftrightarrow$) pointing to the unabbreviated term, leaving it up to the user to decide which form to use. However, in definitions we decided to use abbreviations preferably. In case the dictionary entry is an unabbreviated term, the abbreviation is also indicated after the character S: Example:

8                                                  1   Automatic Control, Systems and Robotics

**SCADA** ↔ supervisory control and data acquisition

**supervisory control and data acquisition** An industrial computer-based system for supervision and control of technological processes. It supports real-time data collection and logging, as well as updates the automatically established values of the variables and parameters of the process. It acts as an HMI and enables the operator to directly interact with devices to check and change process variables and parameters. It can display errors and sound alarms, as well as accept the relevant operator's actions. **S**: SCADA

The same applies to terms composed of an abbreviation and unabbreviated words. Example:

**SIMO system** ↔ single-input multiple-output system

**single-input multiple-output system** A dynamic system, which has one input and several outputs. Such structure enables the development of the controllable canonical form and thus facilitates controller design. **S**: SIMO system

When the arrow points to a term having more than one definition, the preferred term is followed by a number in parentheses indicating to which definition the term refers. Example:

**control feedback** → control loop (2)

**control loop** **1**. A structure, which is designed to maintain a controlled variable at a prescribed set point. **2**. The loop in which the measured controlled signal is compared to the reference signal. A controller uses the difference between the two to generate the control signal. The latter drives the actuator system intending to minimise the difference between the reference signal and the measured controlled signal. **S**: control feedback

# Part II
# Terminological Dictionary of Automatic Control, Systems and Robotics

# Chapter 2
# A

**absolute encoder** **1**. An encoder, which generates a unique code from a system of binary-coded tracks and converts the information about the position relative to a fixed point into the corresponding electrical output, e.g., mechanical encoder, optical encoder, magnetic encoder. It does not lose position information when power is switched off. Its small size and compact construction enable integration in various space-constrained applications. **2**. An encoder, which generates a unique code from a system of binary-coded tracks and converts the information about the orientation of the object relative to a fixed angle into the corresponding electrical output, e.g., mechanical encoder, optical encoder, magnetic encoder. It does not lose orientation information when power is switched off. Its small size and compact construction enable integration in various space-constrained applications.

**absolute maximum** → global maximum

**absolute minimum** → global minimum

**absolute pressure** Pressure, which is measured relative to a perfect vacuum. It is the sum of atmospheric pressure and gauge pressure.

**absolute thermal conductivity** → thermal conductance

**abstract model** The nonphysical presentation of a system, which contains an abstraction of the reality. It can be either a mental model or a symbolic model.

**acausal model** A mathematical model, the equations of which can be given in a neutral form, not considering the computational order. In a simulation, it does not require its input arguments to be explicitly based on previously calculated output arguments. It enables a clear presentation of the system structure and the construction of reusable components in a computer-based modelling tool, e.g., Modelica.

**acceleration-error constant** **1**. A closed-loop system parameter, which is defined by the limit value of the open-loop transfer function multiplied by $s^2$ as $s$ approaches

© ZRC SAZU/Research Centre of the Slovenian Academy of Sciences and Arts 2023
R. Karba et al., *Terminological Dictionary of Automatic Control, Systems and Robotics*,
Intelligent Systems, Control and Automation: Science and Engineering 104,
https://doi.org/10.1007/978-3-031-35755-8_2

0. Here, $s$ is the complex variable in the $s$-plane. **2**. A closed-loop system parameter, the value of which is inversely proportional to the steady-state error of the response of the system to a unit-parabolic signal.

**accelerometer** An electromechanical device, where a correspondingly damped mass on a spring enables the determination of the measured object's acceleration by measuring the spring compression, e.g., piezoelectric accelerometer, MEMS accelerometer, strain gauge accelerometer. It is commonly used in various domains, e.g., engineering, process control, safety control, consumer electronics, biology, medicine.

**accessible workspace** $\rightarrow$ reachable workspace

**accumulation** $\rightarrow$ stock

**accumulator battery** $\rightarrow$ rechargeable battery

**accuracy** **1**. The property of a measuring system where the difference between the measured value and the measurement with no error involved is small. **2**. Proximity of the measured value to the measurement with no error involved, expressing a degree of correctness of a measurement. **3**. The property of a model where the information it provides is correct or near to correct concerning the aim of the modelled system.

**AC/DC converter** $\rightarrow$ rectifier

**Ackermann's formula** **1**. A control-system design method for establishing the parameters of the controller in full-state feedback-control systems. It is used for single-input systems. **2**. A design method for establishing the parameters of a state observer. It is used for single-output systems.

**AC motor** $\leftrightarrow$ alternating-current motor

**across variable** A variable, the value of which is determined by subtracting the values at two end-points of an element, e.g., voltage, linear velocity, angular velocity, pressure, temperature. It is measurable using the parallel connection of the measuring instrument and the corresponding component, e.g., spring, capacitor, storage tank.

**AC tachogenerator** $\rightarrow$ alternating-current tachometer (1, 2, 3)

**AC tachometer** $\leftrightarrow$ alternating-current tachometer (1, 2, 3)

**action connector** $\rightarrow$ link

**actual value** The value of a signal at the considered time instant.

**actuating variable** The variable that excites a change in the behaviour of a dynamic system, frequently used in system identification.

**actuator** 1. A primary element of an actuator system, which amplifies the usually low-energy controller-output signal, using the corresponding power source, and converts it into mechanical motion. It can base on hydraulic, pneumatic, electric, mechanical, thermal or magnetic principles, e.g., electric motor, solenoid, stepper motor, hydraulic motor, pneumatic motor. 2. A single compact device that contains all the elements of the actuator system. 3. A part of a robot, comprising servomotors, drives and transmissions, which carries out either locomotion or manipulation actions by animating mechanical components.

**actuator system** A system, which converts the output signal of a controller into a signal that causes corresponding changes in the controlled-system behaviour. It usually consists of a power amplifier, an actuator, a final control element as well as a signal amplifier and a signal converter. **S**: final control system

**adaptive control** A control system that adjusts to changes in the controlled system. Usually, the adjustment involves varying the parameters of the controller.

**adaptive controller** A controller, which takes into account changes in the dynamics of the controlled system, e.g., parameter adaptive controller, model reference controller.

**adaptivity** The property of a control system to maintain its specified behaviour in spite of changes in the system. The required controlled output is assured by automatically modifying the control-system parameters with regard to changing system parameters.

**A/D converter** ↔ analogue-to-digital converter

**adder** → summator

**adjustable-frequency drive** → variable-speed drive

**admittance control** Control of the deviation of the robot end-effector motion from the desired motion due to the interaction with the environment, which is related to the contact force. The robot and the environment act as two magnets facing together with the same poles. The method is used in collaborative operations and haptic interfaces.

**aerial mobile system** A mobile system, which flies or floats in the air, e.g., aeroplane, helicopter, drone, rocket, missile, satellite, hot-air balloon. It can operate either with or without a pilot.

**affine system** A nonlinear system, which is linear in the input but nonlinear with respect to the state. It is structurally similar to a linear system with an additional constant term.

**agent** A subsystem of hardware or software equipment, which is autonomous, mobile, collaborative, flexible and reactive, e. g., deliberative agent, cognitive agent, hybrid agent. Its possible implementations are an autonomous mobile robot, a software agent, an artificial-life agent or a computer virus.

**agile manufacturing** A quick-response oriented enhancement of lean manufacturing, which ensures rapid reactions to customer needs and market changes, while maintaining cost and quality of products.

**AGV** ↔ automated guided vehicle

**AI** ↔ artificial intelligence

**air cylinder** → pneumatic cylinder

**air motor** → pneumatic motor

**air relay** → pneumatic relay

**alarm panel** A component of an alarm system, which alerts the operator of alarm conditions in the plant. **S**: annunciator panel

**alarm system** A system, which gives an audible, visual or tactile signal about the occurrence of unauthorised, wrong, unforeseen or dangerous situations in a control system.

**algebraic loop** 1. A structure in a block-oriented simulation language, which does not enable the sorting algorithm to finish the procedure. This occurs when the input of at least one block depends on the output of the same block in the same simulation step. Therefore, the simulation model contains a loop without any dynamic blocks. Such a situation requires special treatment, which slows down or even prevents simulation. 2. A structure in an equation-oriented simulation language, which does not enable the sorting algorithm to finish the procedure. This occurs when the same variable appears on the left side and on the right side of at least one equation in the same simulation step. Therefore the mathematical model contains a loop structure without integration or delay. Such a situation requires special treatment, which slows down or even prevents simulation.

**aliasing** A phenomenon, which concerns the frequencies that are above the Nyquist frequency and occurs when reconstructing a continuous-time signal from a discrete-time signal. The high frequencies are mapped to the corresponding lower frequencies. It can be observed when watching an accelerating wheel in a film with a particular frame rate when the wheel seemingly slows down.

**alphatron vacuum gauge** An ionisation gauge, which uses alpha particles to ionise the gas. The number of ions formed on the collector electrode in the chamber results in a voltage directly proportional to the measured quantity. **S**: radioactive ionisation gauge

**alternating-current motor** An electric motor, which consists of stator windings supplied with current, which periodically reverses its direction to produce a rotating magnetic field, and rotor windings, producing a second rotating magnetic field, e.g., synchronous motor, induction motor, single-phase AC motor. Its speed is controlled by the frequency of the current often with a VSD. It is used as a part of, e.g., pump, compressor, fan, machine, robot, actuator system. **S**: AC motor

2 A

**alternating-current tachometer** **1**. A tachometer, which measures angular velocity in one direction, implemented as a precisely constructed generator, which consists of two windings on a stator, one of which is supplied with alternating voltage, and a thin aluminium cup as a rotor connected to the rotating shaft. Eddy currents generated in the rotor induce stator voltage, proportional to the measured angular velocity. It is often used as a feedback sensor in engine-velocity control or in motor-velocity control. **S**: AC tachogenerator, AC tachometer **2**. A tachometer, which measures angular velocity in one direction, implemented as a precisely constructed generator, which consists of a rotor with a permanent magnet connected to the rotating shaft, and stator winding, in which the voltage with the amplitude or frequency, proportional to the measured angular velocity, is induced. **S**: AC tachogenerator, AC tachometer **3**. A sensor, which measures angular acceleration. It is implemented as a precise generator, which consists of two windings on a stator, one of which is supplied with a direct voltage, and a thin aluminium cup as a rotor connected to the accelerated rotating shaft. Eddy currents generated in the rotor induce stator voltage proportional to the measured angular acceleration. **S**: AC tachogenerator, AC tachometer

**ambient sensor** A sensor, which detects various environmental quantities, e.g., light intensity, temperature, humidity, movement, pressure, dust, noise, and possibly recognises faces or pictures. It is primarily intended for smart rooms or for smart buildings and can also be used for environmental monitoring.

**amplification** → gain (1)

**amplitude** The largest difference of the value of a periodic signal from its equilibrium value. **S**: magnitude

**amplitude distortion** The distortion that occurs in an amplifier or other device when the output amplitude is not a linear function of the input amplitude. The measurement conditions must be specified.

**amplitude normalisation** → amplitude scaling (1, 2)

**amplitude response** The part of frequency response, which is the ratio of the output-signal amplitude at a particular frequency and the input-signal amplitude at the same frequency. It is often given in decibels as a part of the Bode plot. **S**: magnitude response

**amplitude scaling** **1**. An algebraic operation, which converts the problem variables of a mathematical model into the corresponding computer variables. For instance, every problem variable is divided by its known or estimated maximum absolute value. Consequently, the obtained computer variable is a dimensionless quantity, which varies between –1 to 1. **S**: amplitude normalisation, magnitude scaling **2**. An algebraic operation that multiplies the amplitude of the signal by a constant value. The procedure is required in analogue simulation but can be helpful also in digital simulation. **S**: amplitude normalisation, magnitude scaling

**analogical model 1**. A model representing a dynamic system, using a corresponding analogy, frequently an electrical analogy. **S**: analogue model **2**. A physical model, represented by another equivalent physical model, which is more comprehensible and enables easier analysis. **S**: analogue model

**analogue computer** A computer, in which physical quantities, e.g., electrical voltage, fluid pressure or mechanical movement, are continuous and analogical to the corresponding quantities of the observed system. An example is a computer, in which mathematical operations, including integration, are carried out with analogue electronic circuits. Due to its calculation speed and parallel structure, it is an efficient tool for the simulation of dynamic systems.

**analogue-digital converter** $\rightarrow$ analogue-to-digital converter

**analogue filter** A passive or active electronic circuit, which processes signals in a frequency-dependent manner. It operates on continuous-time signals and suppresses frequencies that are outside a given frequency range, taking out unwanted noise or signal components, e.g., low-pass filter, band-pass filter, notch filter, Chebyshev filter, Butterworth filter, Bessel filter.

**analogue model** $\rightarrow$ analogical model (1, 2)

**analogue signal** A continuous signal, the amplitude of which can take any value from a continuous interval of possible values.

**analogue signal transmission** An integral part of a control system, which utilises standardised continuous electrical signal or continuous pneumatic signal for the connections and exchange of information among the elements of a control system, e.g., sensors, actuators, analogue electronic controllers, transmitters, transducers, signal converters, amplifiers, filters. The signal can be either voltage, current or compressed air, e.g., 0-10 V DC, 0-20 mA DC, 4-20 mA DC, 20-100 kPa pneumatic signal.

**analogue simulation** A simulation, which enables the experimentation with the mathematical model of a dynamic system that is described with differential equations of various types. The latter are solved by integration, mostly using electrical circuits or other analogies.

**analogue-to-digital converter** A device that converts an analogue signal into the corresponding approximate digital signal. **S**: A/D converter, analogue-digital converter

**analogy 1**. A comparison of diverse components, which are similar in significant aspects, enabling better understanding, clarification or explanation of the problem. **2**. A presentation of diverse processes with the same dynamic properties. It enables the replacement of the considered process components with the equivalent components from another domain, e.g., mechanical components with electrical ones.

**analytical instrument** An instrument, which measures the molecular composition of a material or the concentrations of the material's constituents, e.g., gas chromatograph, HPLC, mass-spectrometer, spectrophotometer, turbidimeter, colourimeter, refractometer. It is used in, e.g., chemical analysis, pharmaceutical analysis, clinical analysis, forensic analysis, food analysis, beverage analysis, petrochemical testing.

**analytical linearisation** → Taylor-series expansion method

**analytical modelling** → first-principles modelling

**android 1.** → humanoid robot **2.** A robot, which is built to resemble a male human.

**anemometer** A sensor, which measures the velocity of atmospheric wind or of a gas flow, e.g., cup anemometer, vane anemometer, hot-wire anemometer.

**aneroid gauge** → capsule

**angle condition** A point in the $s$-plane, in which the angle of an open-loop transfer function equals an odd multiple of $180°$. A corresponding value of the complex variable $s$, representing a closed-loop system pole, lies on the branch of the root locus plot. **S**: angle criterion

**angle criterion** → angle condition

**angle valve** A valve, the inlet valve port and outlet valve port of which are perpendicular to each other, e.g., globe valve, ball valve, check valve. It is commonly used in, e.g., public or residential plumbing, heater drain, boiler feedwater.

**angular acceleration** The time derivative of the angular velocity of a spinning object.

**angular displacement** The change of orientation of a rigid object, which rotates around a fixed axis for a certain angle in a certain direction.

**angular displacement sensor** A displacement sensor, which measures rotary movements, e.g., resolver, rotary encoder, rotary potentiometer. **S**: angular displacement transducer, rotary motion sensor

**angular displacement transducer** → angular displacement sensor

**angular mass** → flywheel

**angular motion** → circular motion

**angular speed** → angular velocity

**angular velocity** The time derivative of the angle of a spinning object. **S**: angular speed, rotary speed, rotational velocity

**animal-inspired robot** → robot animal

18                                                                        2  A

**animal robot** → robot animal

**ANN** ↔ artificial neural network

**annubar flow meter** A differential-pressure flow meter, which measures the average fluid-flow velocity of liquid, gas or steam in pipelines or ducts. It contains multiple pressure tapping ports to obtain the average flow rate, compensating for a nonideal flow profile. It is used in, e.g., chemical industry, petrochemical industry, pharmaceutical industry, gas delivery, liquid delivery. **S**: averaging Pitot tube

**annunciator panel** → alarm panel

**antecedent** The part of an if-then rule, which defines the condition for the follow-up of the statement in the consequence. It is used in a logical statement, as well as in a fuzzy model. **S**: if-part, premise

**anthropocentric approach** A user-friendly strategy of control-system design, which considers human and social aspects.

**anthropomorphic robot** → humanoid robot

**anti-integral windup** → anti-windup

**anti-reset windup** → anti-windup

**anti-windup** The part of a control system, which contains a controller with a specially structured integral term. The integral term is designed to prevent integral windup. In the case of saturation of the final control element or the actuator, the integral term stops the integration or activates a feedback path around the integrator. Therefore, it prevents the nonresponsiveness of the controller or the high overshoot of the controlled signal. **S**: anti-integral windup, anti-reset windup

**aperiodicity border** The value of a parameter in the mathematical model of a system, at which the system response becomes aperiodic. It occurs when critical damping is reached.

**aperiodic response** **1**. A response of a system, which never repeats itself and has an infinite period. It is described with an aperiodic function. **2**. The response of an overdamped system.

**application-area expert** → domain expert

**application-development life cycle** → system-development life cycle

**areometer** → hydrometer

**ARMA model** ↔ autoregressive moving-average model

**ARMAX model** ↔ autoregressive moving-average model with exogenous inputs

**AR model** ↔ autoregressive model

**arrival angle** The angle, by which a branch of the root locus plot enters into a complex conjugate pair of open-loop transfer function zeros, taking into account that the root locus plot is symmetric about the real axis of the $s$-plane.

**articulated robot** A robot, which consists of two or more consecutive rotational joints acting around parallel axes.

**artificial intelligence** A subfield of computer science, which studies machines capable of performing tasks that typically require human intelligence. Many theories, methods and technologies, e.g., visual perception, machine learning, natural language processing, cognitive computing, machine vision, expert systems, speech recognition, handwriting recognition, are included. It is frequently implemented in, e.g., automation, robotics, military, as well as in healthcare, finance, economics and in, e.g., manufacturing robots, drone robots, self-driving cars. **S**: AI

**artificial muscle** An actuator, which mimics a natural muscle, changing its, e.g., stiffness, expansion, reversible contraction, rotation as a result of an external stimulus. It uses various actuating mechanisms, e.g., pneumatic actuation, electric field actuation, electric power actuation, thermal actuation.

**artificial neural network** A structure with topological properties of a network for information processing, frequently used for modelling. It can be classified according to the direction of information flow or according to the optimisation of parameters, e.g., feedforward neural network, recurrent neural network, self-organising map. **S**: ANN

**ARX model** ↔ autoregressive model with exogenous inputs

**assistant robotics** → assistive robotics

**assistive robotics** A subfield of service robotics, in which robots sense, process sensory information and perform actions that benefit seniors and people with disabilities. **S**: assistant robotics

**asymptotic Bode plot** A Bode plot, which is sketched using linear approximations of the sections in the Bode magnitude plot and in the Bode phase plot. The slopes of the approximative lines change at the corresponding cutoff frequencies. It enables a simple determination of the Bode transfer-function form. **S**: straight-line Bode plot

**asymptotic Lyapunov stability** The property of an autonomous system, the solutions of which start out near an equilibrium point in the state space and then converge to the equilibrium point.

**asynchronous motor** → induction motor

**atmospheric pressure** Pressure of the surrounding air, usually measured near the ground. It varies with temperature and altitude.

**attenuation** The reduction of the intensity of a phenomenon, e.g., reduction of the amplitude of a signal.

**autocorrelation** A cross-correlation between the signal and its delayed copy. It is a measure of the repetitiveness of the intervals of a signal and can be used to find repeating patterns, e.g., in a periodic signal. **S**: serial correlation

**auto-manual transfer** Switching from manual control to automatic control and vice versa, which is performed by the operator. Bumpless transfer is often implemented.

**automated guided vehicle** A wheeled robot that is fully automated, custom-made and works without a human operator. It moves in a manufacturing facility or in a warehouse following markers or wires in the floor using, e. g., a vision system, magnets, laser distance sensor. It appears in, e.g., raw-material handling, pallet handling, removal of waste. **S**: AGV, automatic guided vehicle

**automatic control** Control, in which the open-loop-system control variable or the closed-loop-system control variable is determined by a corresponding control system with minimal or reduced human intervention, or completely without it. It is used in various technical and nontechnical applications of any complexity.

**automatic guided vehicle** → automated guided vehicle

**automatic operation** Functioning of a system, process or equipment without human intervention.

**automatic tuning of parameters** → parameter auto-tuning

**automation** **1**. The implementation of automata in technical systems, enabling their self-operation. **2**. The implementation of a control technology in production and service activities, thereby reducing the need for human operation.

**automation pyramid** A schematic representation of different levels of automation in a factory, showing how technology is integrated into the industry. The lowest level is the Field level comprising sensors, actuators and devices. Higher up is the Control level comprising controllers and PLCs. Next is the Supervisory level comprising SCADA and HMI. Higher up is the Planning level with MES. At the top is the Management level with ERP. Additionally, the External level comprises internet, customers, suppliers, collaborators.

**automaton** **1**. A device that operates without human intervention. **2**. Abstract description of a system with a finite number of internal states.

**autonomous jump** A discrete phenomenon in a hybrid system, in which the continuous state changes abruptly as a consequence of the continuous state reaching a certain domain.

**autonomous mobile robot** An autonomous robot, which is not attached to the environment and is able to move around. It is energy independent, makes its own decisions and performs the corresponding actions.

**autonomous mobile system** An autonomous system, which consists of a mechanical part, enabling the required mobility as well as of hardware and software equipment, which is needed for achieving the desired level of autonomy.

**autonomous operation** Independent functioning of a system without any human intervention, often in an unfamiliar environment and at unpredictable circumstances.

**autonomous robot** A robot, which is capable of performing tasks, dealing with the environment and making decisions on its own. It gains information from its surroundings, operates without human intervention or assistance in an extended period of time and avoids situations that are harmful to people, property or itself.

**autonomous switching** A discrete phenomenon in a hybrid system, in which the dynamics of the system changes abruptly as a consequence of the continuous state reaching a certain domain.

**autonomous system  1**. A system which changes its behaviour in response to unanticipated events, e.g., unmanned aircraft, autonomous robot, spacecraft. It is capable of operating in an unknown environment and in unpredictable circumstances without human intervention. Moreover, it can make decisions and implement tasks on the ground, in the air, in the water or in space. During its operation, it is powered by a built-in energy supply. **2**. A system without external excitation, e.g., a system described by an ordinary differential equation without input variable and its derivatives.

**autoregressive model** A regression model, where the output variable is a linear function of its previous values and of a stochastic term, which describes noise. It is used for the prediction of a time series, which defines a discrete-time signal, or as a basic structure for the prediction of the system output, e.g., ARX model, ARMA model, ARMAX model. **S**: AR model

**autoregressive model with exogenous inputs** An AR model of a system, which represents the mapping of the present output value from a finite number of previous values of inputs and outputs as well as from the present value of the input. Its coefficients are optimised according to a one-step-ahead prediction error. **S**: ARX model, autoregressive model with extra inputs

**autoregressive model with extra inputs** → autoregressive model with exogenous inputs

**autoregressive moving-average model** An AR model of a system composed of the regression-model part and the moving-average model part for modelling output noise. It is used for modelling a time series and its one-step-ahead prediction based on the previous values and the present value of the predicted variable. **S**: ARMA model

**autoregressive moving-average model with exogenous inputs** An AR model of a system composed of the regression-model part for modelling input-output relation and the moving-average model part for modelling output noise. **S**: ARMAX model

**auto-tuning control** Control in which the controller parameters are tuned automatically upon request of the operator. It is based on predefined and self-triggered experiments before the controller is put into operation.

**averaging Pitot tube** $\rightarrow$ annubar flow meter

**axial-flow centrifugal pump** A pump, which consists of a rotating impeller in a pipe that causes the increase of the pressure and flow of a fluid with low exploitation of centrifugal force. It is used in, e.g., drainage system, power plant, cooling water supply, sewage handling.

**axial piston pump** A positive displacement pump, which consists of multiple pistons arranged in a rotating housing. The pistons are stroked by a fixed angled plate transferring liquid or gas from the inlet port to the outlet port. It is capable to produce extreme pressure. It is used in, e.g., heavy-duty industry, extrusion, coating, air conditioning.

**azure noise** $\rightarrow$ blue noise

# Chapter 3
# B

**backdrivability** **1**. A control-system design method for establishing the controller parameters in full-state-feedback control systems. It is used for single-input systems. **2**. The ability for interactive transmission between the input axis and the output axis of a transmission system, which means that the transfer function between motor torques and external forces is reversible. **3**. The measure, which determines how accurately the force or movement produced at the output of a transmission system is transferred to its input.

**backlash** A phenomenon in a transport mechanism caused by gaps between its parts, which at the change of the direction of movement expresses an oblique hysteresis shape of the static characteristic. It appears in, e.g., gearbox, actuator, belt coupling.

**back-pressure proximity sensor** → pneumatic proximity sensor

**backpropagation** A deterministic parameter-optimisation method, which uses the gradient of an objective function, e.g., for ANN training.

**backstepping** A method for designing stabilising control for a special recursive class of nonlinear dynamic systems. The procedure begins with stabilised irreducible subsystem and proceeds with the design of controllers, which stabilise each outer subsystem to achieve complete-system stability despite uncertainties in system parameters and disturbances.

**backward elimination** An iterative method for the selection of input variables of a mathematical model of a system, where the variables are systematically removed from the model. The elimination of each variable is tested with a chosen model-fit criterion.

**backward rectangular rule** A frequency-response fitting method for the system discretisation, which approximates the derivative of a variable with the difference between its value in the actual time step and its value in the previous time step, divided by sampling time.

© ZRC SAZU/Research Centre of the Slovenian Academy of Sciences and Arts 2023
R. Karba et al., *Terminological Dictionary of Automatic Control, Systems and Robotics*, Intelligent Systems, Control and Automation: Science and Engineering 104, https://doi.org/10.1007/978-3-031-35755-8_3

24                                                                    3  B

**balance law**  → conservation law

**balancing feedback**  → negative feedback

**ball-shaped robot**  → spherical robot (1)

**ball valve**  A valve, which uses a hollow, perforated pivoting sphere to control the flow of fluid in a pipe. A hole through the centre of the sphere usually has equal diameter as the inner diameter of the pipe, that results in a small pressure drop in the fully open position. It can be driven manually, electrically or pneumatically and is mostly used as a quick-acting on-off valve and less commonly as a control valve, due to its limited flow-control accuracy. It is used in, e.g., industrial plant piping, household piping.

**bandwidth**  A band of frequencies, at which the output of a system tracks input signal accurately, enabling a fast response of the system frequencies, for which the closed-loop amplitude response does not drop below $-$ 3 dB, e.g., the frequencies below the cutoff frequency of a low-pass filter.

**bang-bang control**  → on-off control

**bang-bang controller**  → on-off controller

**bank**  → roll angle

**barometer**  A sensor, which measures atmospheric pressure, e.g., mercury barometer, aneroid barometer, MEMS pressure sensor. It is used in, e.g., car engine, smartphone, as well as for, e.g., weather forecast, altitude measurement.

**base coordinate frame**  A Cartesian coordinate frame attached to the robot base. Its $z$-axis points out of the base and is perpendicular to it.

**base quantity**  A quantity comprised in the International System of Quantities, which are length ($l$), mass ($m$), time ($t$), thermodynamic temperature ($T$), electric current ($I$), amount of substance ($n$), luminous intensity ($I_v$). All other quantities can be expressed in terms of these seven quantities. See Table 1. **S**: fundamental quantity

**base unit**  The unit of a base quantity, where the corresponding symbol is either metre (m), kilogram (kg), second (s), kelvin (K), ampere (A), mole (mol) or candela (cd). See Table 1. **S**: fundamental unit

**basic control**  The lower level of the process-control level of the CIM with the aim of achieving and maintaining the desired state of the technological-process variables or the desired states of its subsystems, e.g., valve manipulation for substance dosage in the food industry. It enables the fulfilment of the procedural-control goals.

**basic physical dimension**  The expression of the considered quantity with the product of the symbols for the dimension of the base physical quantities L, M, T, $\Theta$, I, N, J. Their exponents can be positive, negative or 0, e.g., the quantity dimension of acceleration is denoted $LT^{-2}$. See Table 1. **S**: fundamental physical dimension

**Table 1** Base quantity, base unit, basic physical dimension

| Base quantity | Base-quantity symbol | Basic physical dimension | Base unit | Base-unit symbol |
| --- | --- | --- | --- | --- |
| Length | $l$ | L | metre | m |
| Mass | $m$ | M | kilogram | kg |
| Time | $t$ | T | second | s |
| Thermodynamic temperature | $T$ | $\Theta$ | kelvin | K |
| Electric current | $I$ | I | ampere | A |
| Amount of substance | $n$ | N | mole | mol |
| Luminous intensity | $I_V$ | J | candela | cd |

**basis function** A function from a set of elementary functions used for the approximation of an arbitrary function, e.g., a radial basis function, a ridge basis function.

**batch process** A process of production, the mass or energy flows of which are interrupted due to separate stages of charging, content processing and discharging of system units, e.g., blast furnace, rotary kiln, chemical reactor. **S:** batch system

**batch system** $\rightarrow$ batch process

**behavioural validity** The model-validation procedure, which asseses the similarity of the dynamics of the system and its model with the same input signals.

**bellows gas flow meter** A gas flow meter, which consists of two flexible chambers, the volumes of which can be changed by expansion or compression. The gas flow is directed by internal valves, which alternately fill and empty the chambers, producing a nearly continuous flow of fixed volumes of gas towards the output. The movements of the chambers are caused by the pressure difference between the inlet and outlet and are transformed into rotations that are counted.

**bellows pressure sensor** An absolute pressure sensor or differential pressure sensor, which consists of a container that expands as a response to the force applied by the pressure within, creating a displacement proportional to the measured pressure. The container is usually a seamless, flexible, thin-walled metallic tube that has deep folds to enable expansion and contraction. It is suitable for measuring low pressure or medium pressure.

**bias** **1.** The property of the expected value of a parameter estimate, which differs from the true value of the parameter. **2.** The property of a statistical method, which causes the expected value of a parameter estimate to differ from the true value of the parameter. **3.** A value, which describes the difference between the average of measurements performed on the same object and the measurement with no error involved, e.g., the difference between laboratory average value and reference-laboratory average value for the same measurements of the same quantity at the same object.

**BIBO stability** ↔ bounded-input bounded-output stability

**bidirectional elimination** An iterative method for the selection of input variables of a mathematical model of a system that is a combination of forward selection and backward elimination. The addition or elimination of each variable is tested with a chosen model-fit criterion.

**bilinear system** A nonlinear system represented by a differential equation with two variables. The differential equation is linear in relation to each variable, but nonlinear when both are taken into account.

**bilinear transformation** A frequency-response fitting method for the system discretisation, which approximates the natural logarithmic function with a first-order approximation that is an exact mapping of the $z$-plane to the $s$-plane. **S**: Tustin's method

**bilinear transformation with pre-warping** A modified bilinear transformation, in which frequency warping is compensated at a selected frequency.

**bimetal thermometer** A temperature sensor, which converts temperature changes into mechanical displacements, e.g., spiral strip bimetal thermometer, helical strip bimetal thermometer. Two materials with different temperature coefficients are welded in a strip. Thermal expansion of the strip results in a deflection, which is proportional to the measured temperature.

**biofeedback** A technique of making unconscious or involuntary bodily processes, e.g., breathing, heart rate, blood pressure, brain wave, muscle tension, stress, to become perceptible to human senses by electronic monitoring with, e.g., electromyograph, electroencephalograph, electrocardiograph, electrodermograph, pneumograph. It enables human training to achieve conscious mental control of some problematic bodily functions.

**bio-inspired robotics** → biologically-inspired robotics

**biologically-inspired robotics** A subfield of robotics, which studies biological systems, looking for the mechanisms that may solve some specific engineering problems in a simple and effective way. Examples are various types of robotic snakes, robotic fishes, and multilegged robots. **S**: bio-inspired robotics, nature-inspired robotics

**biorobotics** A subfield of robotics, which combines cybernetics, bionics and genetic engineering in the design of robots that emulate living organisms mechanically or even chemically, e.g., artificially created organisms, the use of biological organism as a component of a robot.

**biosensor** An analytical device, which consists of a sensitive biological element, e.g., tissue, cell receptor, enzyme, and a transducer, e.g., optical transducer, piezoelectric transducer, electrochemical transducer. The transducer transforms the measured signal into some usable, usually electric, form, which is shown on a corresponding display. It is used in, e.g., clinical application, diagnostic application, environmental application, industrial application.

**biproper transfer function** A proper transfer function, in which the degree of the numerator is equal to the degree of the denominator. Its relative degree equals 0.

**black-box model** A mathematical model of a dynamic system obtained from measurements of the input signals and output signals, e.g., ARX model, ARMAX model, FIR model.

**black noise** Complete silence or mostly silence with bits of random noise.

**blending function** → membership function

**block** An element of a block diagram, usually depicted as a rectangle, which represents the unidirectional functional connection between its input and its output.

**block diagram** The graphical representation of system structure and signal connections, which consists of blocks that contain mathematical models of subsystems, directed signal paths, summing points and branch points. **S**: block scheme

**block-diagram algebra** The set of rules, which enables the transformation of a block diagram to an equivalent form, e.g., simplification of the block diagram, structural changes of the block diagram.

**block-diagram reduction** The procedure, which enables the presentation of a block diagram in the simplified equivalent form using the block-diagram algebra.

**blocking zero** → transmission zero

**block-oriented simulation language** A simulation language with a graphical preprocessor, which enables the simulation of mathematical models that are given in the form of an explicit block structure. The user connects the corresponding entities from a library of predeveloped blocks and defines their parameters. In such a way, rapid simulation-scheme development is enabled.

**block scheme** → block diagram

**blue noise** Coloured noise, the power spectral density of which is directly proportional to the frequency. Therefore, its power spectral density increases by 10 dB/ decade. **S**: azure noise

**BMS** ↔ building management system

**Bode amplitude plot** → Bode magnitude plot

**Bode diagram** → Bode plot

**Bode magnitude plot** The part of a Bode plot with logarithmic frequency scale along the $x$-axis and amplitude in decibels on the $y$-axis. **S**: Bode amplitude plot

**Bode phase plot** The part of a Bode plot with logarithmic frequency scale along the $x$-axis and phase in degrees or radians on the $y$-axis.

**Bode plot** A graphical representation of the frequency response, which is used for analysis and control-system design. It consists of the Bode magnitude plot and the Bode phase plot that are shown on separate graphs, both with the same logarithmic frequency scale along the $x$-axis. **S**: Bode diagram

**Bode transfer-function form** A modified zero-pole-gain transfer-function form, which enables direct determination of the steady-state gain, as well as the time constants of the system, being the reciprocal of its poles and zeros. **S**: time-constant transfer-function form

**bond graph** A graphically presented physical-system model, in which the energy-flow relations and connections among the elements of a dynamic system are described by two independent parameters, i.e., effort and flow. Its noncausal character enables the reuse of submodels of multidomain systems.

**bounded-input bounded-output stability** A property of a system, the output of which is always bounded if the input is bounded. **S**: BIBO stability, input-output stability

**Bourdon tube** An absolute pressure sensor, which consists of either a C-shaped or a helical metal tube with an elliptic cross-section and one end sealed. When the pressured medium enters the tube it tends to straighten the C-shaped tube or extend the helical tube. The deflection is proportional to the measured pressure.

**Bourdon tube thermometer** → pressure spring thermometer

**Box-Jenkins model** A mathematical model, comprised of the input-output model part and the noise-model part, which are mutually independent.

**branch** 1. A line segment in a signal-flow graph or in a block diagram, which connects two nodes. It has its direction and gain and is symbolised by an arrow-ended line. The input variable represents the cause, while the output variable is the consequence, but not vice-versa. **2**. The part of the root locus plot that shows all possible positions of a single closed-loop pole as the consequence of the variation of a certain system parameter, which is often the gain. The number of such segments equals the number of the open-loop-transfer-function poles.

**branch point** An element of a block diagram or of a signal-flow graph, depicted as a dot on a signal connection, which enables simultaneous usage of the same signal in different locations. **S**: pickoff point, takeoff point

3  B                                                                    29

**breakaway point**  The point on the real axis or the complex-conjugate pair of points in the $s$-plane, in which a branch in the root-locus plot is split into several branches due to multiple roots.

**break frequency**  $\rightarrow$ cutoff frequency (1, 2)

**breakpoint**  $\rightarrow$ cutoff frequency (1, 2)

**Bridgman gauge**  An absolute pressure sensor, which measures extremely high static and dynamic pressures. It consists of a fine manganin wire that is wound in a coil, which is enclosed in a corresponding kerosene-filled pressure container. A Wheatstone bridge measures the resistance of the manganin wire, which changes under pressure.

**Brownian noise**  Coloured noise, the power spectral density of which is inversely proportional to the square of the frequency. Therefore, its power spectral density decreases by 20 dB/decade. **S**: Brown noise, red noise

**Brown noise**  $\rightarrow$ Brownian noise

**brushless DC motor**  A DC motor with permanent magnets on the rotor and a fixed armature on the stator. The electrical connection to the moving armature is therefore not required. Electronic controller, according to the rotor position measurements, continuously switches the phase of the stator windings to keep the motor touring. It is used in, e.g., electrical vehicle, cordless tool, drone, heating, air conditioning, industrial engineering, actuator system. **S**: electronically commutated motor

**brute force**  A fundamental problem-solving technique, which systematically enumerates all possible candidates for the solution and checks if they meet the conditions. It finds a solution if it exists, but is mostly applicable to limited-size problems where the simplicity of implementation is more important than the speed of problem-solving. **S**: exhaustive search, explicit enumeration, generate and test

**bubbler level sensor**  A level sensor, which consists of a tube immersed in a liquid and a system that monitors and controls the flow rate of purge gas, often air, through the tube. The flow rate of the gas in the dip tube is kept constant with the varying downstream pressure caused by variations in the liquid level. This pressure is equal to the hydrostatic pressure of the liquid at the submerged end of the tube where the gas exits. The hydrostatic pressure is proportional to the measured level of liquid with a known density.

**building automation system**  $\rightarrow$ building management system

**building management system**  Supervisory system used for monitoring and the control of mechanical and electrical equipment in a residential, commercial or industrial building, e.g., heating, ventilation, air conditioning, lighting, power system, alarm system. **S**: BMS, building automation system

**bumper sensor** A switch, which reports a mobile robot or vehicle if it collides with an object. If it has whiskers, the collision with an obstacle is detected before it happens. It enables automatic avoidance of crashes. It is typically used in a robotic vacuum cleaner.

**bumpless transfer** A feature of a control system that prevents abrupt changes of the control signal. It is implemented for switching between automatic control and manual control or between different control signals.

**business planning and logistics** → business-planning level

**business-planning level** The top level of the CIM, which comprises business processes and company management. **S**: business planning and logistics

**butterfly valve** A valve, which influences the flow of fluid by rotating a disc, which is positioned in the centre of the pipe. A rod passing through the disc enables manual operation with, e.g., handles, gears, or automatic operation with, e.g., electric actuator, pneumatic actuator, hydraulic actuator. It is used in, e.g., water supply, wastewater treatment, gas handling, fuel handling, as well as in chemical industry, pharmaceutical industry, food industry.

# Chapter 4
# C

**CACE** → computer-aided control-system design

**CACSD** ↔ computer-aided control-system design

**CAD** ↔ computer-aided design

**CAE** ↔ computer-aided engineering

**calculation interval** → step size

**calibration** 1. A process of evaluating and adjusting the precision and accuracy of a measuring instrument by comparing its values with the values of a traceable reference instrument of known accuracy at several points along the scale of the instrument. 2. A process of adjusting the measuring instrument to meet the specifications of the manufacturer.

**CAM** ↔ computer-aided manufacturing

**CAN** ↔ controller area network

**canonical form** The simplest possible standardised mathematical model of a linear SISO system that ensures no loss of information, e.g., diagonal canonical form, controllable canonical form, observable canonical form, Jordan canonical form. The transformation of a mathematical model into such an equivalent form might simplify some methods of analysis and synthesis of control systems significantly. In the case of a linear MIMO system, the system matrix is given in the corresponding block form **S**: normal form, standard form

**capacitive angle sensor** An angular displacement sensor, constructed as a capacitor, which measures circular movement by changing the effective area of its plates.

---

© ZRC SAZU/Research Centre of the Slovenian Academy of Sciences and Arts 2023
R. Karba et al., *Terminological Dictionary of Automatic Control, Systems and Robotics*, Intelligent Systems, Control and Automation: Science and Engineering 104, https://doi.org/10.1007/978-3-031-35755-8_4

**capacitive displacement sensor** 1. A displacement sensor, constructed as a capacitor, which exploits the influence of the measured movement on the area of its plates, on the distance between its plates, or on the dielectricity between its plates. 2. A noncontact movement sensor, distance sensor or position sensor, which can also be used for thickness measurements. It can achieve resolutions in the submicrometre range and can operate in extreme temperatures.

**capacitive humidity sensor** A humidity sensor, which utilises the dependence of the dielectricity of a hygroscopic material on its water content. A thin strip of metal oxide or hygroscopic polymer film is placed between two electrodes forming a capacitor, the capacitance of which changes when the dielectric material absorbs water. The changes, that are proportional to the relative humidity are measured either directly or through the frequency of the oscillator that includes the capacitor with the hygroscopic dielectric. It is used in, e.g., moisture-sensitive industries, food processing, air conditioning, weather stations, household appliances.

**capacitive level sensor** A level sensor, which consists of two electrodes or one electrode and the metal wall of the container as the second electrode. The measured impedance between the two electrodes is affected by either the dielectricity of nonconductive fluid with a known and stable dielectric constant or by the effective area of the electrodes that depends on the level of conductive fluid. It is used in applications dealing with different types of fluids, e.g., liquid metal at very high temperature, dissolved gas at very low temperature, very high-density medium.

**capacitive linear encoder** A linear encoder, which senses the change of modulation of a high-frequency signal transmitted between the comb-shaped slider and the fixed stator. The contact via electrostatic field enables accurate linear-displacement measurement.

**capacitive proximity sensor** A proximity sensor, constructed as a capacitor, which detects the presence of a metallic or a nonmetallic object, the dielectricity of which is different from air dielectricity.

**capillary viscometer** A viscometer, which consists of a U-shaped glass tube with two bulbs, one higher and one lower, connected by a tube with a hairlike internal diameter, e.g., U-shaped viscometer, gas capillary viscometer. The measured viscosity is proportional to the time of passage of a known quantity of fluid through the connection between the bulbs considering the known geometry of the device.

**capsule** An absolute pressure sensor, which consists of two circular-shaped diaphragms welded at the edge enabling the measured pressure to act from the inside on both diaphragms simultaneously. Compared to a single diaphragm, it causes twice the displacement that is proportional to the applied pressure. A stack can be used to further increase the displacement. It is especially suitable for measuring relatively low pressure of gaseous media and is therefore often used as a barometer. **S**: aneroid gauge

4 C

**Cardan angles** → Tait-Bryan angles

**Cartesian robot** A robot with three controlled axes, which is made up of three prismatic joints, the axes of which are perpendicular to each other. Therefore, its reachable workspace has a prismatic shape. It is often used as a part of, e.g., CNC machine, 3D printer, pick-and-place, plotter. **S**: linear robot, rectangular robot

**cascade control** A control structure with a master control loop and a nested slave control loop. The output of the master controller provides the setpoint for the slave controller. It is used when more than one measurement but only one control variable is available, enabling better control performance. The slave control loop must be faster than the master control loop.

**CASE** ↔ computer-aided software engineering

**causal FOH** → delayed FOH

**causal loop diagram** A graphical representation of qualitative influences among treated quantities in a system. It is the result of the first step when using SD as a modelling method. **S**: CLD

**causal model** A mathematical model or a part of it, which is described with an input-output model, defining the output as a consequence of the input of the system. In a certain block of a block-oriented simulation language, e.g., ACSL, Simulink, the input is unidirectionally converted to the output. It is a quantitative representation of the dynamics of the system enabling the prediction of the modelled system behaviour.

**causal system** A system, the output values of which depend only on past input values and current input values, but not on future input values, e.g., a system described by a proper transfer function. It is physically realisable as its response does not precede its excitation.

**cavitation** A phenomenon, in which vapour bubbles, created by valve flashing, collapse and implode when liquid pressure increases after passing an obstacle or contraction. It causes severe damage, e.g., corrosion, noise, vibration, making the inner surfaces of the valve rough and covered with tiny pits. It may also affect piping, instruments and other equipment, significantly shortening the lifespan of the device.

**CC-Link** An open industrial Ethernet, which enables the communication of devices from numerous manufacturers, e.g., CC-link IE, CC-link LT, CC-link Safety. It is used in, e.g., manufacturing industry, process control, building automation, facilities management.

**CC-Link IE** ↔ CC-Link industrial Ethernet

**CC-Link industrial Ethernet** A CC-Link with gigabaud data throughput for connecting devices at the controller level and for connecting to the enterprise level, e.g., CC-link Field, CC-link Control. **S**: CC-Link IE

34  4 C

**cellular automaton** A mathematical model, which consists of cells, arranged in a grid. It evolves in parallel at discrete time steps, changing cell states according to a defined set of simple rules, that includes also the states of neighbouring cells. The rules govern the iterative replication and destruction of cells. It is used as a model of a variety of complex discrete-event systems and continuous systems composed of simple units in, e.g., physics, chemistry, biology, cryptography, epidemiology, parallel computing.

**centralised control** A control structure where one computer, controller or PLC controls several control loops.

**centre of mass** The average position of all body parts of a system, e.g., a humanoid robot, pondered with the mass of its elements. **S**: COM

**centre point** An equilibrium point of the second-order system in the phase plane, where all trajectories are ellipses and the singularity point is their centre.

**centroid** The point on the real axis of the $s$-plane, in which the asymptotes of a root locus plot intersect. When the difference between the number of open-loop transfer function poles and the number of open-loop zeros is greater than 0, it represents the number of zeros with infinite value and consequently the number of asymptotes. The angles of the asymptotes against the real axis of the $s$-plane can also be calculated. **S**: intersection point

**channel** $\rightarrow$ path (1)

**chaotic system** A nonlinear dynamic system described by a mathematical model with explicitly defined constant parameters. Its behaviour is very dependent on the initial values. It does not converge to an equilibrium point, to a limit cycle or any other quasi-periodic dynamics.

**characteristic equation** **1**. An equation, which is obtained by setting the denominator polynomial of a system transfer function to 0. It can be derived also from a differential equation or state equations. Its roots completely determine the stability of a linear SISO system. **2**. An equation, which is obtained by setting the characteristic polynomial of the closed-loop transfer function to 0.

**characteristic polynomial** **1**. A polynomial associated with a matrix, which gives information about that matrix. It is closely related to the determinant of the matrix and its roots are the eigenvalues of the matrix. **2**. A polynomial, which is expressed as the sum of one and the open-loop transfer function. It is used in the stability analysis of a closed-loop system.

**characteristic response** $\rightarrow$ natural response

**chattering** Unwanted mechanical vibrations in the contacts of relays, switches or contactors, caused by imperfect contact between two contact surfaces due to, e.g., ageing effects, corrosion, burned contact points, too low or too high voltage. It can cause signal distortion in an analogue or digital electronic circuit as well as in a control loop. **S**: contact bounce

4 C

**check valve** A valve, which allows the flow of liquid or gas in only one direction causing minimal pressure drop. It is usually used for the protection against the reverse influence of media, i.e., backflow. **S**: clack valve, counterflow valve, nonreturn valve, one-way valve, reflux valve, retention valve

**chilled-mirror dew-point humidity sensor** A humidity sensor, which directly measures the dew-point temperature of a gas. A sample of gas with unknown water vapour content condenses on an inert cooled reflective metal surface. A beam of light is reflected from the surface into a photodetector enabling the detection of the condensed fine water droplets. A control system maintains the metal surface temperature, at which condensation equals evaporation. This defines the dew-point temperature that is in turn calibrated to the relative humidity of the measured gas. Due to its accuracy, it is often used as a standard in metrology laboratories as well as in moisture-sensitive industries.

**choked flow** A condition, in which the flow rate through a valve does not change substantially despite a pressure-drop variation. For gas, it occurs when the flow velocity reaches the speed of sound in that gas. For liquid, it occurs with the onset of valve flashing, where the flow passage is fully cluttered with vapour bubbles. Therefore, the theoretical nominal flow through a certain valve cannot be reached, which must be taken into account in valve sizing. **S**: critical flow

**chromatograph** An analytical instrument, which separates, identifies and quantifies each component in volatile, temperature-stable compounds that can be vapourised or dissolved in a liquid solvent, e.g., planar chromatograph, column chromatograph, gas chromatograph, HPLC. A mixture is passed through a medium, in which the components move at different rates. The mobile phase is an inert or unreactive carrier gas or solvent, while the stationary phase is a microscopic layer of liquid or polymer on inert solid support inside a metal or glass tube-shaped column or on a plate. A corresponding detector enables the identification and quantification of the components in a sample.

**CIM** ↔ computer-integrated manufacturing

**circular motion** The movement of a rigid object, which rotates around an axis. **S**: angular motion, rotational motion

**clack valve** → check valve

**classic Euler angles** → proper Euler angles

**CLD** ↔ causal loop diagram

**closed kinematic chain** A kinematic chain, the elements of which are connected in parallel.

**closed-loop control** → feedback control

**closed-loop system** A system, the signal-flow graph or block diagram of which consists of a forward path and one or more feedback paths.

**closed-loop transfer function** A transfer function, which dynamically relates the controlled signal to the reference signal.

**closed system** A physical system, which is not influenced by external forces and does not exchange matter with its surroundings. Therefore, no mass or energy transfer in or out of the system occurs. **S**: isolated system

**cluster analysis** → clustering

**clustering** An unsupervised learning, which finds a structure in a set of data using different algorithms. It organises entities into groups, the members of which are similar according to predefined metrics. It is commonly used in, e.g., optimisation, identification, pattern recognition, data mining, machine learning. **S**: cluster analysis

**μC** → microcontroller

**CMG** ↔ control moment gyroscope

**CMM** ↔ coordinate measuring machine (1, 2)

**CNC** ↔ computer numerical control

**CNC machine** A flexible programmable device, which consists of a machining tool, e.g., drill, lathe, mill, plasma cutter, laser cutter, 3D printer, and a controller containing a program with the detailed description of steps in the processing procedure. It enables fully automated small-scale production.

**cobot** → collaborative robot

**coder** **1**. A device, which converts information into the desired form. **2**. A device that generates computer software without human intervention. **3**. A person who writes computer software according to specifications.

**collaborative operation** Robot operation, which combines the precision, power and endurance of a robot with the capability of a human operator for solving problems. The human operator works in direct proximity to the robot system, while it is active. Physical contact between the human operator and the robot may occur within the collaborative workspace.

**collaborative robot** A robot, which is used in collaborative operation with a human operator. Therefore, the safety features must be secured through the lightweight design of the robot, flexible actuators, various sensors and advanced control. **S**: cobot, cooperative robot

**collaborative workspace** The part of the reachable workspace, in which a robot system and a human operator perform tasks concurrently during operation. Therefore, strict limitations concerning speed, space limits, and torque sensing are applied to guarantee the safety of the human operator.

## 4 C

**collision avoidance 1.** → obstacle avoidance **2.** The strategy of an autonomous system that enables the design of the modified trajectory when an obstacle is detected.

**coloured noise** A noise with a specifically shaped power spectral density. Unlike white noise, its power spectral density is not constant.

**colourimeter** A photometer that measures how much light of a certain wavelength is transmitted through the measured substance, which selectively transmits only one wavelength, while the others are absorbed or reflected. The transmitted light is measured by a corresponding photodetector enabling the determination of the concentration of a particular substance in the sample. The light is filtered to obtain a required one-wavelength light. Its calibration is based on the comparison with the known sample. It is used for monitoring, e.g., growth of yeast or bacteria, water quality, plant nutrients, course of chemical reactions, substandard or counterfeit drugs, haemoglobin in the blood.

**COM** ↔ centre of mass

**combinational control** Logic control, which uses only basic logic functions. It maps the actual inputs of the controller to outputs, without considering any previous events or signals. **S**: combinatorial control

**combinational logic** A mathematical formalism, which describes combinational control that can be implemented using only Boolean elements. **S**: combinatorial logic, time-independent logic

**combinatorial control** → combinational control

**combinatorial logic** → combinational logic

**combined simulation 1.** A simulation, which merges continuous simulation and discrete-event simulation, e.g., a simulation of a batch process. **2.** A simulation of a system, which is described by differential equations on the whole observation interval or on a part of it. However, at least one state variable or its derivative is not continuous over simulation time, e.g., a simulation of a system containing a hysteresis.

**combined simulation language** A simulation language, which combines the capabilities of a continuous simulation language and of a discrete-event simulation language. It includes mechanisms for the transition from continuous simulation to discrete-event simulation and vice versa, as well as algorithms for the numerically accurate treatment of discontinuities. Thus, simultaneous simulation of continuous dynamics and discrete events is enabled.

**communication interval** An interval between the moments, in which the numerical results in a digital simulation are available.

**communication protocol** A set of rules, which supports data exchange among two or more devices in a network, e.g., internet protocol, transmission control protocol, hypertext transfer protocol, PROFIBUS protocol, HART protocol, PROFINET protocol. It is implemented in hardware and software using serial data transmission or parallel data transmission. It includes syntax, semantics, synchronisation of digital communications and analogue communications as well as possible authentication, error detection and error correction.

**compact PLC** A PLC, which consists of a fixed number of inputs and outputs, a central processing unit and the power supply, all comprised in one housing. It is typically designed to perform basic functions and to control less demanding systems with a limited number of signals. S: fixed PLC, integrated PLC, unitary PLC

**companion form** A canonical form of a state-space model, in which the bottom row or the rightmost column of the system matrix contains the negative coefficients of the characteristic polynomial, while the rest of the system matrix is a unit matrix. S: Frobenius canonical form, Frobenius normal form, phase variable form, rational canonical form

**compartmental model** → compartment model

**compartment model** A linear or nonlinear mathematical model, which consists of a finite number of submodels. The submodels are assumed to be homogeneous entities with lumped parameters that are connected mutually as well as with the environment. It is often depicted in graphic form and used in, e.g., biopharmacy, biomedicine, ecology, modelling of chemical reactions, system theory. S: compartmental model

**compensation** A design method for modifying the open-loop transfer function by complementing the existing dynamic structure. Consequently, it changes the closed-loop behaviour of the system, according to steady-state requirements and prescribed transient-response specifications, e.g., by cancelling poles and zeros, by modifying the frequency response, by altering the root locus.

**compensator** A controller, which implements compensation for modifying the open-loop transfer function to achieve the required closed-loop performance. Its design considers steady-state requirements, transient-response specifications, and frequency-response requirements.

**compiler-oriented simulation language** A simulation language, which translates the source program into a higher-level general-purpose language using the corresponding compiler. Simulation is quick and flexible, while the program is easily portable.

**complementary root locus plot** A root locus plot, in which the negative control-loop gain is the variable parameter.

**complementary sensitivity function** The transfer function between the setpoint and the controlled variable in a closed-loop system. It is used in the robustness analysis of the control system for setpoint tracking.

4   C

**complex programmable logic device**   PLD, which contains programmable function blocks connected by a global reconfigurable interconnection matrix, enabling the generation of complex logic components. It is often used in high-performance control applications. Due to its small size and low power consumption, it is suitable for cost-sensitive, battery-operated portable applications. **S**: CPLD

**compliance**   **1**. The property of a robot, which allows small displacements due to elastic behaviour between the robot end-effector and the robot gripper or tool. **2**. → flexibility (1)

**compliance control**   Admittance control, in which only the static relationship between the robot end-effector pose deviation from the desired motion and the contact force or contact torque is considered.

**componental design**   → modular design

**compressed-air system**   A system, which powers a corresponding distribution network that delivers the compressed air to users with minimal leakage and minimal loss of pressure. It consists of one or several compressors, conditioning equipment, e.g., aftercooler, separator, filter, air dryer, and air distribution equipment, e.g., valving, air conditioning, pressure control.

**compressor**   A power source, which increases the pressure of gas by reducing its volume, e.g., reciprocating compressor, rotary vane compressor, rotary screw compressor, diaphragm compressor, centrifugal compressor. It enables the transport of gas or air through a pipe. It is often used in an actuator system of diverse pneumatic systems, e.g., heating, ventilation, air conditioning.

**computational intelligence**   → soft computing

**computational model**   → computer model

**computed torque control**   → inverse dynamics control

**computer-aided control engineering**   → computer-aided control-system design

**computer-aided control-system design**   The application of dedicated computer software for the analysis, design and testing of control systems. **S**: CACE, CACSD, computer-aided control engineering

**computer-aided design**   A computer technology, which helps engineers to build models and 2D or 3D drawings throughout the design, construction and operation of, e.g., processing plants, factories. It enables modifications, analyses or optimisation of a design procedure and is easy to edit and easy to share. It is used in, e.g., automotive industry, aerospace industry, ship-building industry. **S**: CAD

**computer-aided engineering** A process of solving engineering problems by the use of sophisticated interactive graphical software complemented with CAD and CAM. It enables simulation, validation and optimisation of processes, products or manufacturing, as well as performance insight in earlier phases of process development. It is used in, e.g., control-systems analysis, multibody-dynamics analysis, reliability analysis, robustness analysis. **S**: CAE

**computer-aided manufacturing** An application technology, which uses software, computer-driven machinery, real-time control and robotics to facilitate and automate the operation of a manufacturing plant. It is often upgraded with CAD and enables increased production speed, more precise tooling accuracy, ease of customisation and improved material usage. It is used in, e.g., milling, drilling, sawing, welding, flame-cutting, water-cutting, plasma-cutting, laser-cutting, pick-and-place. **S**: CAM

**computer-aided software engineering** Application of computer-assisted tools and methods ensuring enhanced-quality and error-free software. It merges methodologies and software equipment, which supports the designer primarily in the development phases of, among others, control systems. The approach covers the whole cycle of product development including, e.g., code generation, repositories, prototyping. **S**: CASE, computer-assisted software engineering

**computer-assisted software engineering** → computer-aided software engineering

**computer control** An activity, in which a computational device, e.g., microcomputer controller, microcontroller, PLC, PAC, is used to direct the operations in the manufacturing industry and in the process industry.

**computer-integrated manufacturing** A form of manufacturing supported by computers that control the entire production process. It is based on the feedback paradigm and integrates the production and business processes, thereby allowing individual processes to exchange information with each other. It combines various technologies, e.g., CNC, CAD, CAM, CAE, ERP, techniques, e.g., feedback control, supervisory control, and equipment, e.g., sensors, actuators, controllers, PLCs. It usually has a hierarchical structure and consists of a business-planing level at the top, a production-control level in the middle and a process-control level at the bottom. **S**: CIM, flexible design and manufacturing

**computer model** A computer program that attempts to simulate a mathematical model of a particular process or plant on a digital, analogue or hybrid computer. In such a way, efficient experimentation with the corresponding simulation model is enabled. **S**: computational model

**computer numerical control** Computer control of a machining tool, which enables completely automated material processing as well as a direct connection to the CAD or CAM software tools. **S**: CNC

**computer simulation** The process of imitating the behaviour of a real system by experimenting with a mathematical model. It is a simplified representation of the system behaviour, usually performed on a digital computer, or optionally on an analogue computer or a hybrid computer.

**computer vision** An interdisciplinary technology, which acquires, processes, analyses and interprets digital images or videos with various sensors, e.g., digital camera, TIC, tactile sensor. Therefore, information about the structure, shape or topology of the observed object or scene can be obtained. The results are applicable in QC as well as in control and supervision of diverse systems.

**concurrency** A property of data processing that multiple computations are carried out at the same time.

**conditional stability** The property of a linear time-invariant system that it is stable for a set of values of a particular parameter, and unstable for a different set of values of the same parameter, e.g., when the gain of the system is changed.

**condition/event net** → Petri net

**condition number** The number defined as a ratio between the largest and the smallest singular value of the transfer function matrix, which is consequently frequency dependent.

**conductivity level sensor** A level sensor, which is used for point level detection and for level switching of conductive liquids. The border between the conductive and nonconductive liquid or gas can be detected. The liquid closes the alternating-current circuit between two or multiple electrodes or between the metal container wall and the electrode. A sudden increase of current generates a switching signal indicating the measured level point or points.

**connectivity** A property of the assembly between two robot links in a kinematic chain described by the number of variables. It must be specified at a given time instant to enable the complete description of the relative pose between the two links.

**connector** → link

**consequence** The part of an if-then rule, which defines the effect of the trueness of the statement in the antecedent. It is used in a logical statement, as well as in a fuzzy model. **S**: then-part

**conservation law** A fundamental natural law, which states that some quantity or property of an isolated physical system remains constant as the system evolves over time, e.g., energy balance, mass balance, momentum balance. Consequently, the considered quantity can be neither created nor destroyed, though it can appear in different forms. **S**: balance law

**consistency** The property of a parameter estimation that the sequence of estimates converges in probability to the true value as the number of data points used for the estimation increases indefinitely.

**constrained optimisation method** An optimisation method, which considers the corresponding explicit or implicit boundaries on the optimised parameter values, e.g., linear programming, quadratic programming, nonlinear programming, penalty method.

**constructability** The property of a system, the present state of which can be determined from the present and past outputs and the inputs of the system in finite time. Any observable system is also a constructable one.

**contact bounce** $\rightarrow$ chattering

**contact model** A model of forces and allowed relative motions, which are transmitted through the physical touch, resulting in friction and possible deformation. It is determined by the geometry and the material properties of the touch surfaces.

**contact sensor** A sensor, which detects and measures the collision between a robot end-effector and its environment, e.g., tactile sensor, force-torque sensor.

**continuous controller** $\rightarrow$ continuous-time controller

**continuous process** A process in which matter, energy or information are uninterruptedly flowing through a system.

**continuous simulation** A simulation of a system, described by differential equations of various types. It can be implemented as digital simulation or analogue simulation.

**continuous simulation language** A simulation language, which enables the simulation of mathematical models that are described with ordinary linear or nonlinear differential equations as well as with partial differential equations.

**continuous system** A production system, the mass or energy flows of which are uninterrupted, e.g., heat exchanger, plug flow reactor, continuous stirred-tank reactor.

**continuous-time control** The control, in which the controlled signal is measured uninterruptedly. Additionally, the control signal is a continuous function of the input signal of the controller.

**continuous-time controller** A controller that calculates its continuous output signal from the continuous input signal. **S**: continuous controller

**continuous-time model** A mathematical model with dependent variables, which are defined on a continuous range of time, described with, e.g., ordinary differential equations, partial differential equations. **S**: continuous-variable model

**continuous-time signal** **1**. A signal that is uninterrupted in the observed time interval. **2**. The output signal of a D/A converter.

**continuous-time system** A system, which is described with a continuous-time model. Thus, a continuous-time input signal produces a continuous-time output signal.

**continuous-variable model** → continuous-time model

**control** **1**. A procedure, which affects the operation of the system through selected actions to achieve the prescribed goal, e.g., by changing quantities in the system, by determining the order of operations. **2**. A procedure, which enables the transformation of information about a system and its environment into decisions and actions. Considering the given criteria and constraints, the desired goals can be achieved.

**control algorithm** A formal representation of the operation steps for automatic control, e.g., mathematical representation, pseudocode, flowchart.

**control canonical form** → controllable canonical form

**control chart** A statistical tool, used in QC to analyse and understand process variables, to determine process capabilities, to monitor the effects of the variables on the difference between the target and the actual performance, to predict future performance and to estimate the sources of variations. It graphically presents one quality indicator versus time or sample number. Its allowable deviation, defined with upper and lower control limit, is also presented, enabling the excessive variations of the variable to be detected. **S**: process-behaviour chart

**control design** A multistep procedure, which uses control theory or experimental approach to achieve the fulfilment of the specified requirements for the controlled system behaviour. It enables the synthesis of the control strategy and of the controller structure as well as the determination of the parameter settings. It can be achieved using several methods, considering the properties of the system to be controlled and the control objectives, e.g., stability, safety, environmental rules, technological limitations, economic factors, efficient management of technological systems as well as of nontechnological systems.

**control engineering** A multidisciplinary field, which designs, analyses, optimises, implements, maintains and manages control systems to assure the desired behaviour of the controlled system.

**control equipment** Devices and corresponding software, which enable control of a system, e.g., PLC, sensor, actuator, communication device.

**control feedback** → control loop (2)

**control horizon** A parameter in MPC denoting the number of future time steps, in which the control variable is allowed to change when seeking the optimal control-variable sequence in a particular time step. It is less than or equal to the prediction horizon.

**controllability** The property of a system, for which a control signal exists such that the zero-value state can be reached in a finite time from any initial state. For a continuous-time linear system, it is equivalent to reachability. A corresponding discrete-time system may not exhibit reachability.

**controllable canonical form** A canonical form of a state-space model, which ensures the controllability of the system. It can be obtained, e.g., with the transformation of the transfer function into the simulation scheme using the nested method. The description enables a direct determination of the characteristic polynomial and transfer function of the system from the bottom row of the obtained system matrix. **S**: control canonical form, controller canonical form

**control law** A mathematical expression describing the controller output in dependence on the controller input, which enables the transformation of the error signal to the control signal.

**controlled jump** A discrete phenomenon in a hybrid system, in which the continuous state changes abruptly as a consequence of control action.

**controlled signal** The signal at the output of a controlled system. **S**: manipulated signal

**controlled switching** A discrete phenomenon in a hybrid system, in which the dynamics of the system changes abruptly as a consequence of control action.

**controlled variable** The output of a closed-loop system or of an open-loop system, which is intended to be as close as possible to the setpoint. **S**: manipulated variable

**controller** **1**. A system, which adjusts the behaviour of a dynamic system by influencing the input of the dynamic system, e.g., microcomputer controller, analogue electronic controller, mechanical controller, pneumatic controller. **S**: regulator **2**. A device, which detects mostly binary inputs and generates outputs according to its internal rules, e.g., PLD, PLC, relay circuit, digital circuit.

**controller adjustment** → controller tuning

**controller area network** A serial network protocol, which is used for high-speed time-critical embedded communication between a microcontroller and devices without the need for a host computer. To connect a device, an appropriate receiver and a transmitter are required. It is frequently used in automotive applications, as well as applications in harsh environments, e.g., vehicle electronic networking, process control, industrial automation, agricultural equipment, medical equipment, building automation. **S**: CAN

**controller canonical form** → controllable canonical form

**controller parametrisation** → controller tuning

**controller tuning** A procedure, which enables the determination of adjustable controller parameters. Its goal is to assure the best possible closed-loop performance of the controlled system. Different approaches can be used, e.g., tuning rules, simulation, trial-and-error method. **S**: controller adjustment, controller parametrisation

**control loop** 1. A structure, which is designed to maintain a controlled variable at a prescribed set point. 2. The loop in which the measured controlled signal is compared to the reference signal. A controller uses the difference between the two to generate the control signal. The latter drives the actuator system intending to minimise the difference between the reference signal and the measured controlled signal. **S**: control feedback

**control matrix** → input matrix

**control moment gyroscope** A mechanism made up of a flywheel and one or more motorised gimbals, which is used to influence the orientation of a spacecraft by changing its axis of rotation and thus generating torque. Therefore, it replaces thrusters or external applications of torque. **S**: CMG

**control signal** 1. The signal that is generated at the controller output and implements the control action by driving the actuator system. **S**: manipulative signal 2. → link

**control system** A system, which manages, commands, directs or regulates other devices or systems to achieve or maintain the desired response using control loops. It is mostly computerised and implemented as a set of mechanical or electronical components.

**control theory** An interdisciplinary engineering discipline, which deals with the behaviour of dynamic systems by designing the way of influencing a particular physical system so that it behaves according to certain desired specifications. It includes system analysis and synthesis of controllers that enable the controlled system to achieve and maintain the proposed dynamics.

**control valve** A valve, which changes the flow rate of the fluid according to the corresponding control signal, e.g., globe valve, gate valve, ball valve. It consists of a valve actuator, a valve positioner and a valve body enabling continuous setting of all states from completely opened to completely closed. It may provide the additional possibility of manual intervention. It is the most common final control element in process industry.

**control variable** The input of the controlled system in a closed-loop system or an open-loop system, which influences the controlled variable through the dynamic properties of the controlled system. **S**: manipulative variable

**control vector** → input vector (1)

**converter** 1. A device, which transforms matter, energy or information from one form to another form. 2. → dynamic variable

**cooperative manipulation** Manipulation of a common object by two or more robots.

**cooperative robot** → collaborative robot

**coordinate frame** A coordinate system, with regard to which the pose of a body is defined.

**coordinate measuring machine** **1**. A device, which measures the geometry of a physical object with a corresponding sensor, e.g., tactile sensor, optical sensor, laser sensor, proximity sensor. It can be controlled either manually or by a computer. In manufacturing and assembly processes, it measures the workpieces according to the design goal. **S**: CMM **2**. A passive robotic mechanism with a sensor at the robot end-effector, enabling contact or contactless assessment of distance. **S**: CMM

**coriolis density meter** A density meter, which consists of two U-shaped tubes filled with the measured fluid, a driving mechanism that oscillates the tubes at their natural frequency, and a corresponding frequency detector. The change in the fluid density influences the mass flowing through the tubes and in turn changes the natural frequency, which is proportional to the measured density. It can measure a variety of fluids under different operating conditions.

**coriolis mass flow meter** A mass flow meter, which consists of one or more U-shaped tubes energised by constant-frequency vibrations. When liquid or gas passes through the tube, mass-flow momentum causes a change in vibration and twist of the tubes. Two sensors detect the oscillation in time and space. The obtained phase shift is proportional to the measured mass flow rate.

**corner frequency** → cutoff frequency (1, 2)

**cost function** → loss function

**counterflow valve** → check valve

**CPLD** ↔ complex programmable logic device

**crisp set** A set, a particular element of which can be either its full member or not its member at all. An element cannot partly belong to it.

**criterion** An objective function used for quantitative measuring of the performance of a control system, e.g., integral criterion. **S**: performance index

**criterion function** → objective function

**critical damping** Damping, where the step response of a system reaches the new steady state as quickly as possible without an overshoot. In such a case, the damping factor is 1.

**critical flow** → choked flow

**critical gain** → ultimate gain

**critical period** → ultimate period

**cross-correlation** A measure of the similarity of two signals. It is used to find repeating patterns, e.g., in a periodic signal, and for finding a short known signal in a longer signal, e.g., in pattern recognition. **S**: sliding dot product, sliding inner product

**cross coupling** A connection between different input-output pairs of a MIMO system where the control loops interact with each other, which often complicates the control design. It is represented by the corresponding off-diagonal element of the TFM. **S**: interaction

**cross-validation** The method for model validation that assesses how the model predictions generalise to values of input signals, which have not been used for modelling priorly.

**current estimator** A system that is used for determining the states of a treated system from the measured input and output signals. It uses a suitable gain matrix, which considers both the previous measurements as well as the actual output measurement, thereby influencing the dynamics of the estimation error.

**current-to-pressure converter** An analogue electrical transmission, which converts a standardised current signal from measurement or control instrumentation to an equivalent proportional standardised pneumatic signal. **S**: i/p converter

**curve fitting** The procedure, which enables the changes in the structure or parameters of a mathematical model in such a way that the model response fits the measured data points or the prescribed curve as much as possible. **S**: model fitting

**cut-and-try** → trial and error

**cutoff frequency** **1**. The frequency, at which the amplitude response of a low-pass dynamic system is 3 dB lower than at frequency 0. It defines the bandwidth of the dynamic system. **S**: break frequency, breakpoint, corner frequency **2**. The frequency at the intersection of two asymptotes in an asymptotic Bode plot, which is the reciprocal of the corresponding time-constant value in the Bode transfer-function form. It is the point in the frequency response of a system at which the energy flow through the system begins to attenuate. **S**: break frequency, breakpoint, corner frequency

**cybernetics** A transdisciplinary science of communication and automatic control in natural and man-made systems. It connects various domains, e.g., control systems, mechanical engineering, electrical network theory, computer science, communication, evolutionary biology, neuroscience, anthropology, psychology. It is relevant for, e.g., mechanical systems, physical systems, biological systems, cognitive systems, computational systems, social systems.

**cylindrical robot** A robot, which has one rotational joint and two prismatic joints. Therefore, its tube-shaped reachable workspace extends along the axis of the rotational joint and has an annular cross-section in any plane perpendicular to that axis. It is often used in, e.g., assembly operations, handling of machine tools, spot welding.

# Chapter 5
# D

**D/A converter** ↔ digital-to-analogue converter

**Dall-tube flow meter** A differential-pressure flow meter, where the restriction in a pipe is generated by the correspondingly shaped insertion in a tube.

**damped natural frequency** The frequency of the damped oscillatory time response of the dynamic system, which is the result of the previous excitation. The time response oscillates with decreasing amplitude.

**damped oscillation** Oscillatory time response, which is the consequence of natural or artificial damping, caused by energy loss in the system, in turn causing the amplitude of oscillation to decrease with time. Its damping factor is greater than 0.

**damping 1**. A decrease of the oscillation amplitude of a dynamic system, e.g., electrical system, mechanical system. **2**. An action, which causes the oscillation of a system to be reduced, restricted or prevented. It is the consequence of different processes that dissipate the energy stored in oscillation, e.g., air friction, liquid friction, eddy current. **3**. A property of a measuring system where the oscillation amplitude of its output decreases and stabilises at a steady state in a reasonably short time.

**damping coefficient** → damping factor

**damping factor** A positive and dimensionless value that is a measure of the decay of oscillations in a system. **S**: damping coefficient, damping ratio

**damping ratio** → damping factor

**data adjustment** → data cleansing

**data cleaning** → data cleansing

© ZRC SAZU/Research Centre of the Slovenian Academy of Sciences and Arts 2023
R. Karba et al., *Terminological Dictionary of Automatic Control, Systems and Robotics*,
Intelligent Systems, Control and Automation: Science and Engineering 104,
https://doi.org/10.1007/978-3-031-35755-8_5

50        5   D

**data cleansing** The procedure, which enables the process of detecting and correcting of incomplete, incorrect, inaccurate or irrelevant data. Various methods can be applied, e.g., grouping, moving average, interpolation, smoothing. **S**: data adjustment, data cleaning, data editing, data preprocessing

**data editing** → data cleansing

**data mining** A process of collecting, searching-through and analysing a large preexisting database to find useful patterns, trends or relationships using the corresponding software. It combines tools from statistics, artificial intelligence, machine learning, and database management. It is frequently used in, e.g., process automation, surveillance, business, biology, medicine.

**data preprocessing** → data cleansing

**DC gain** → steady-state gain

**DC motor** ↔ direct-current motor

**D controller** ↔ derivative controller

**DC tachogenerator** → direct-current tachometer

**DC tachometer** ↔ direct-current tachometer

**DDM** ↔ direct digital manufacturing

**dead band 1**. A property of a system where a change of its input signal, which is small enough, does not affect its output signal. **S**: dead zone **2**. A region where the changes of the input signal in any direction do not produce any change in the output of a measurement system or in the control signal. **S**: dead zone

**dead-beat controller** A discrete-time controller, which enables the controlled system to reach a steady state in a finite number of steps. The smallest possible number of steps is equal to the order of the controlled discrete-time system.

**dead reckoning** Navigation, which estimates the current pose of a mobile robot from the known previous pose and the measured relative displacements from the previous pose using, e.g., odometry, inertial navigation. **S**: deduced reckoning

**dead time 1**. The interval of time between the start of an input change and the start of a system response. **S**: time delay, transportation lag, transport delay **2**. A parameter determined by the difference between the start of the step-input change and the intersection of the tangent at the inflexion point of a reaction curve with the steady state before the step-input change. It is used for tuning a PID controller in, e.g., step-response method. See Figure 1. **S**: delay time (2)

**dead zone** → dead band (1, 2)

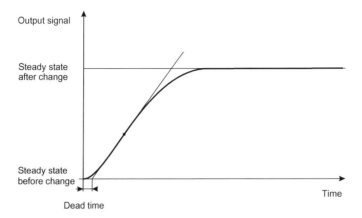

**Fig. 1** Dead time (2)

**decade** 1. A range of amplitude from a particular value to ten times that value. 2. A unit for measuring a frequency ratio of ten, usually between two frequencies on a logarithmic scale, defining the frequency range between frequency and its tenfold value. It is used for expressing slopes of asymptotes in the Bode magnitude plot and the Bode phase plot of the asymptotic Bode plot.

**decentralised control** 1. The control of large-scale systems with local controllers for lower-dimensional subsystems. Such an approach reduces the problems with obtaining and storing data, as well as questions concerning the possible geographical separation of subsystems. The closed-loop structure enables the use of simple controllers and computationally efficient control. 2. The control of MIMO systems with SISO controllers in each path. The influence of the cross-couplings can reduce the control efficiency, so it is frequently impossible to meet all the design requirements.

**decoder** An electrical device that converts a coded signal into the corresponding meaningful signal.

**decoupled system** A linear MIMO system with compensated cross-couplings that is described by a diagonal TFM. As a consequence, any input-output pair is completely independent of other pairs. Hence, it can be regarded as a set of mutually independent SISO subsystems. **S**: diagonal system, non-interacting system, non-interactive system

**decoupling** A MIMO control-design metodology that intends to compensate for the cross-couplings in the system, enabling the design of SISO controllers for the obtained mutually independent SISO subsystems.

**decoupling zero** The zero of a linear MIMO system, representing an uncontrollable and unobservable system mode that does not appear in the TFM, e.g., input decoupling zero, output decoupling zero.

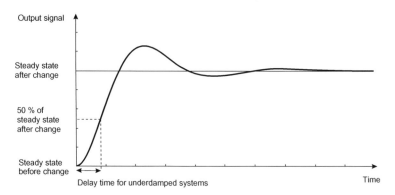

**Fig. 2** Delay time (1)

**deduced reckoning** → dead reckoning

**deductive reasoning** The reasoning about specific conclusions on the basis of general statements, It is often used in testing hypotheses or theories.

**defuzzification** The operation, which assigns a numerical value to a variable according to its membership values appurtenant to each fuzzy set.

**degenerative feedback** → negative feedback

**degree of freedom 1.** The number of design parameters that can be arbitrarily and independently modified, e.g., pole-assignment design parameter. **S**: DOF **2**. The number of subsystems in a control-system structure that can be used and configured independently, e.g., a two-degree-of-freedom controller with two compensators, one in the forward path and one in the feedback path. **S**: DOF **3**. The number of variables in a mathematical model, which has to be defined to completely describe the state, status or pose of the system. **S**: DOF **4.** → robot mobility

degree of fulfilment → membership value

**delayed FOH** A FOH that postpones the signal by one time step. It linearly interpolates between the sampled value in the previous sampling instant and the sampled value in the actual sampling instant. As both values are available online, it is a causal conversion. **S**: causal FOH

**delay time 1.** The transient-response specification that defines the time, in which the transient response of a proportional underdamped second-order system to a unit step signal reaches 50 % of the steady-state value for the first time. See Figure 2. **2.** → dead time (2)

**deliberative agent** A software agent, which processes images of the external environment to generate a map that enables planning of actions and their realisation. Its behaviour is reasonably sophisticated, frequently using the most popular belief-desire-intention architecture. It is often used in multi-agent simulations. **S**: intentional agent

**Delta robot** A parallel robot, the hexagonal platform of which is connected with its hexagonal base by three lateral, parallelogram, leg-shaped mechanisms. Each has one motorised rotational joint, two passive spherical joints and two passive universal joints. It is lightweight and allows for extreme dynamic performance.

**Denavit-Hartenberg notation** A 4x4 transformation matrix, which describes the pose of a coordinate frame with respect to another coordinate frame with four parameters, namely with two distances and two angles.

**densimeter** → density meter

**density meter** A sensor, which measures the mass per unit volume of fluid stored in a tank or flowing in a pipe, e.g., pycnometer, hydrometer, oscillating U-tube, Coriolis density meter, ultrasonic density meter, nuclear density gauge. It is designed as a tabletop, handheld or portable instrument. It is used in, e.g., mining, oil and gas production, paper industry, wastewater treatment. **S**: densimeter

**departure angle** The angle, by which a branch of the root locus plot exits from a complex conjugate pair of open-loop transfer function poles, taking into account that the root locus plot is symmetric about the real axis of the $s$-plane.

**derivative-action time** → derivative time

**derivative control** A control strategy, in which the controller output is proportional to the rate of change of the error. Hence, the controller output is proportional to the time derivative of the error. It rapidly responds to error change and has no direct impact on the steady-state error. **S**: differential control

**derivative controller** A controller, the output of which is the rate of change of the error multiplied by a constant. Its input is the error, whereas its output is the control signal. It can speed up the control-system response and make it more stable. On the other hand, it can not improve a steady-state error and amplifies noise signals. Therefore, it is never used alone. **S**: D controller

**derivative-free method** An optimisation method, which does not require gradient calculations or gradient approximations and uses only availability of the objective function, e.g., direct-search method, PSO, genetic algorithm, simulated annealing, Nelder-Mead method.

**derivative gain** A constant, which is the factor between controller output and the rate of change of the error. It is used in feedback control with the controller that contains a D term. It is given also as the product of proportional gain and derivative time.

**derivative term** A subsystem, the output of which is obtained by differentiating its input signal and multiplying the result by a constant. Usually, a first-order system is implemented to filter the result. It is usually a constituent part of a controller. **S**: differential term, D term

**derivative time** A coefficient, which determines the impact of the D term of a controller on closed-loop performance. It is equal to the time, in which the share of the D term is equalised with the share of the P term in the step response of a PD controller. **S**: derivative-action time, pre-act time

**derived quantity** A quantity, which is defined in terms of the seven base quantities. It can be expressed also as the product of the symbols of the base quantities.Their exponents can be positive, negative or 0.

**derived unit** The unit of a nonbase quantity. It is dimensionless or a product of base units with the corresponding exponents. Its symbols are, e.g., newton (N), pascal (Pa), joule (J), unit of acceleration (m s$^{-2}$).

**describing function** The ratio between the Fourier coefficient of the first harmonic component of the output signal and the amplitude of the sinus signal at the system input. The other coefficients of the Fourier series are neglected. It is an approximation of a transfer function of a nonlinear system and can be used in the analysis of certain nonlinear control problems based on quasi-linearisation.

**describing-function method** A linearisation that provides the approximative equivalent of nonlinear system dynamics in the frequency domain using describing function. **S**: harmonic linearisation method

**descriptive realism** A property of the model that bases on the correct or probable assumptions about the mechanisms in the modelled system. In such a manner, the structure of the model is evaluated. **S**: descriptive reality

**descriptive reality** $\rightarrow$ descriptive realism

**design** A component of the engineering process, which defines a series of steps used in creating functional products or processes that satisfy the specified requirements. It results in new facilities of a system, as well as in expansion or modification of existing facilities. Its multidisciplinary and iterative character enables diverse implementations, e.g., control design, CAD, engineering design, industrial design, process design, path planning.

**detectability** The property of a system that has all the unstable states observable and all the unobservable states stable. It is a weaker property than observability. It can be ensured despite some unobservable states.

**deterministic model** A mathematical model, the output of which is precisely determined by the previous and current inputs. Here, no random relations among variables are taken into account. Such a model always produces the same output signal for a given input signal.

**deterministic optimisation method** An optimisation method that generates and uses deterministic variables, e.g., gradient-optimisation method, linear programming, quadratic programming, nonlinear programming.

**deterministic system** A system, which does not involve any randomness in the procedure of the future states generation. It is modelled by a deterministic model.

**deviation model** A mathematical model, which is described with input deviation variables and output deviation variables. It improves the description clarity of the dynamic changes in the time response of the system.

**deviation variable** A variable defined as the difference between the current value and the value in the operating point, i.e., the coordinate-system origin is shifted to the operating point.

**device language message specification/companion specification for energy metering** A data collection protocol for smart energy metering, control and management, which includes object-oriented data models, application layer protocol and media-specification communication profiles. It is used in various utility-service applications for different kinds of energy, applying any kind of communication media. **S**: DLMS/COSEM

**dexterity** The ability to move the robot end-effector or to apply forces and torques in arbitrary directions.

**dexterous workspace** The area, which can be reached by the arbitrarily-oriented robot end-effector.

**DFT** $\leftrightarrow$ discrete Fourier transform

**diagonal canonical form** A canonical form of a state-space model obtained using the corresponding transformation matrix, which consists of linearly independent eigenvectors of the system matrix. It converts the system matrix to a diagonal matrix, which has unequal eigenvalues as its nonzero elements. Such a description enables the determination of the possible uncontrollable and unobservable states. **S**: modal canonical form, spectral canonical form

**diagonal dominance** The property of the mathematical model of a stable linear MIMO system. In the TFM of such a system, the absolute value of the individual diagonal element must be greater than the sum of the absolute values of the non-diagonal elements in the corresponding row or column for the chosen frequency, i.e., row diagonal dominance, column diagonal dominance.

**diagonal system** $\rightarrow$ decoupled system

**diaphragm pressure sensor** A pressure sensor, which consists of a chamber divided with a circular metal or fibrous flexible disc that is either flat or has concentric corrugations. Its deflection, affected by the pressure of gaseous, liquid and possibly aggressive media, is proportional to the measured pressure. It is used for measuring relatively low pressure. **S**: membrane pressure sensor

56    5  D

**diaphragm pump** A positive displacement pump, which uses the reciprocating action of a rubber or thermoplastic membrane and corresponding on-off valves. Due to its versatility, it can be used in every industry that requires a fluid transfer, e.g., cleaning, spraying, dewatering, filling, dispensing, chemical dosing. S: membrane pump

**diaphragm valve** A valve, which consists of a valve body, an elastomeric membrane and a stem pushing the membrane towards the seat. Its working parts are isolated from the fluid. Therefore, it is also usable in corrosive and abrasive applications, e.g., treatment of slurries, minerals processing, as well as in pharmaceutical industry, chemical industry, food industry. S: membrane valve

**diesel generator** A diesel-fueled engine with compression ignition, which powers an electric generator to provide electrical power. It is often used as a power supply of an actuator system.

**differential control** → derivative control

**differential drive** A drive mechanism of a wheeled robot, which consists of two drive wheels mounted on a common axis. Angular velocity of each wheel is independently controlled by a separate motor. By varying the relative angular velocities of wheels the desired trajectory of a robot can be achieved.

**differential-equation solver** A computer program that implements an appropriate numerical integration method for simulating a dynamic system described by a corresponding mathematical model, e.g., state-space model.

**differential pressure** Pressure obtained by subtracting one of the applied absolute pressures from the other.

**differential-pressure flow meter** A flow meter, which determines the volumetric flow rate from a pressure-drop measurement, e.g., orifice-plate flow meter, flow nozzle, Venturi tube, Dall tube, elbow flow meter. The pressure drop is caused by a partial restriction of the flowstream in a pipe. Two pressures are measured before and after the obstruction. It has a simple construction without moving parts. It is used in various industries where water-flow, oil-flow, gas-flow or steam-flow measurements are needed.

**differential-pressure level sensor** A level sensor, which measures the liquid height above the installation point of the pressure sensor in an open tank, or the height between the high-pressure and low-pressure connections to the vessel in a closed tank. The pressure is proportional to the measured level. It is suitable for measuring, e.g., level of suspended liquid, level of high viscous medium, level in a sealed pressure vessel, level at extreme temperature, level under special hygienic requirements.

**differential system** → differentiating system

**differential term** → derivative term

## 5 D

**differentiated white noise** → violet noise

**differentiating system** A system, the dynamics of which is described by a transfer function with one or more zeros in the origin of the *s*-plane, e.g., a system where a step signal at the input causes the output to settle at the initial value. **S**: differential system

**differentiator** 1. A device or circuit, the output of which is a signal proportional to the rate of change of the input signal. 2. A block of a simulation scheme, which generates an output equal to the first derivative of the input.

**diffuse proximity sensor** A photoelectric proximity sensor or ultrasonic proximity sensor, which consists of an emitter that produces light beam or sound pulses, and a receiver in the same module. The target object that acts as a reflector bounces the light beam or sound pulses back to the receiver. The module is installed in the corresponding location and the target object passes the light beam or sound pulses. The receiver is reached by a light beam or sound pulses that are different from the emitted ones causing the reaction of the sensor output.

**digital-analogue converter** → digital-to-analogue converter

**digital communication channel** A system, which consists of devices and signal connections, enabling the transmission of digital information from source to user.

**digital computer** An electronic computer, which performs calculations and logical operations with numbers, symbols or quantities represented with digits, usually in binary notation. It processes information as a sequence of operations according to instructions that are stored in its memory. It can be used for, e.g., control of the industrial process, simulation of the behaviour of the dynamic system, analysis and organisation of data.

**digital control** Control that is implemented on a discrete-time-calculation-based system, such as digital computer, microcontroller, PLD.

**digital filter** A discrete-time system, which is used to perform mathematical operations on a discrete-time signal, reducing or enhancing certain aspects of that signal. It is a discrete-time equivalent of the analogue filter. It usually consists of DSP with corresponding software as well as of A/D converter and D/A converter enabling the connection with the process.

**digital prototyping** → virtual prototyping

**digital signal** A discrete-time signal with only one value from a finite set of values of amplitude in each time sample.

**digital signal processor** A special-purpose microprocessor chip, the architecture of which is optimised according to the needs of processing signal data in real time. It is used in, e.g., audio signal processing, digital image processing, telecommunications, speech recognition and can be embedded in, e.g., radar, sonar, digital camera, smartphone, high-definition television. **S**: DSP

58 5 D

**digital signal transmission** An integral part of a control system, which utilises standardised digital signal for the connections and exchange of information among the elements of a control system e.g., discrete sensors, on-off actuators, PLCs, microcomputer controllers, microcontrollers, converters, computers. The digital signal voltage is, e.g., 24 V DC, 230 V AC.

**digital simulation** A simulation, which enables the experimentation with the mathematical model of a dynamic system that is described with differential equations of various types. The latter are solved by consecutive integration using numerical integration algorithms.

**digital simulation system** A system, which enables simulation on a general-purpose computer, e.g., simulation language, simulation package.

**digital-to-analogue converter** A device that converts a digital signal into the corresponding analogue signal. **S**: D/A converter, digital-analogue converter

**digital twin** A simulation model with a graphical 3D virtual representation of a physical product, process or system. It collects real data about the system, facilitates model simulation and reflects the current conditions of the asset in operation. It integrates IIoT, artificial intelligence, machine learning into software representation, which enables creating, testing, building, diagnosing and predicting in a virtual environment. It brings several advantages, e.g., increased reliability and availability, reduced risk, lower maintenance cost, improved production, shorter time-to-value.

**dimensional analysis 1.** A procedure, which tests the correctness of the mathematical model by inserting the base or derived units in the equations. The dimensional consistency on both sides of each equation shows their regularity, while discrepancies might detect possible faults or sometimes determine even their locations. **S**: factor-label method, unit factor method **2.** An analysis of relationships among different physical quantities presented with their base quantities and base units. They enable diverse calculations, conversions and comparisons, while the property of dimensional homogeneity is used as the plausibility test for different mathematical relations. **S**: factor-label method, unit factor method

**dimensional homogeneity** A principle, which allows only physical quantities with the same unit to be compared, equated, added or subtracted. However, the quantities with different units can be multiplied or divided.

**dimensionless quantity** A quantity, whose basic physical dimension is 1. Hence, its unit is not explicitly shown, e.g., Reynolds number, Mach number, atomic weight. **S**: pure number, quantity of dimension 1

**Dirac delta function** A function defined on a continuous domain that is 0 everywhere but at the origin of the coordinate system, where it is infinitely high. Its integral equals 1. **S**: unit impulse function

**direct-acting control** $\rightarrow$ direct action

**direct acting control valve** A control valve, which tends to be opened when the flow increases. It requires a special construction of the valve actuator, as well as of the valve disc and the valve seat.

**direct action** The mode of operation of an industrial controller, in which the decrease of the measured controlled variable causes the increase of the control variable. For instance, if the output temperature of the system falls below the reference temperature, which increases the error, the power of the heater must increase. Consequently, negative feedback is achieved, enabling a stable closed-loop system. **S**: direct-acting control

**direct-current motor** An electric motor, which consists of armature windings on the rotor, stationary windings for field poles on the stator, and a rotor-mounted commutator as a rotary switch that enables rotations of the rotor, e.g., shunt DC motor, series DC motor, compound DC motor, brushless DC motor. A direct proportionality of the motor speed to armature voltage and inverse proportionality to the magnetic flux produced on-field poles enable simple speed control. **S**: DC motor

**direct-current tachometer** A tachometer, which measures angular velocity in both directions of rotation. It is implemented as a precisely constructed generator, which consists of a permanent-magnet stator, and a rotor with a set of windings. The rotor is connected to the rotating shaft on one side and to a commutator with brushes on the other side to produce direct-voltage output. The induced voltage is proportional to the measured angular velocity. It is often used as a feedback sensor in engine-velocity control and motor-velocity control of, e.g., conveyor, fan, mixer, machine tool. **S**: DC tachogenerator, DC tachometer

**direct digital manufacturing** A manufacturing paradigm, which enables the production of objects directly from a CAD file or data using 3D-printing technologies, e.g., layered-object manufacturing, stereolithography, selective laser sintering. It eliminates investments in tooling and reduces the time lag between design and production. It is commonly used in, e.g., low-volume production, manufacturing of replacement parts, prototyping, production of special-edition parts. **S**: DDM

**direct dynamic model** A mathematical model, which enables the calculation of robot-end-effector trajectory from known robot-joint forces and torques. **S**: forward dynamic model

**directional control valve** → spool valve

**direct kinematics** Kinematics, in which the robot end-effector pose is calculated from the known robot-joint variables. A similar calculation can also be carried out for robot end-effector velocity and acceleration. **S**: forward kinematics

**direct matrix** → feedforward matrix

**direct method** A method, which enables straightforward derivation of the simulation scheme from a mathematical model described by an algebraic equation or a formula.

**direct-Nyquist-array method** The MNA method that attenuates cross-couplings according to the Nyquist-plot graphical criterion for the evaluation of diagonal dominance. The corresponding MIMO controller can be implemented with simple and commonly used control components. The obtained system is considered to be decoupled. Therefore, the design of additional SISO controllers for the individual subsystems is necessary. **S**: DNA method

**direct problem** The problem of determining the output signal from the known input signal, dynamics and initial conditions of the system. Such a problem is often solved using simulation.

**direct-search method** A derivative-free method for solving nonlinear optimisation problems. It relies exclusively on the values of the objective function, searching a set of points around the current point and comparing each trial's solution with the best previously obtained solution to proceed towards the final solution. It is often used in system analysis and design due to its simplicity, flexibility, robustness and reliability. **S**: pattern-search method

**discrete actuator** → on-off actuator

**discrete controller** → discrete-time controller

**discrete event** A momentary occurrence of the change of the system state, which happens sequentially at a certain time instant.

**discrete-event model** A mathematical model, for which the observation interval of time is divided into nonequidistant subintervals, determined by events, that are defined by the correspondingly conditioned dependent variable.

**discrete-event simulation** A simulation, the states of which are changing according to the occurrence of events in discrete instances. The events in discrete instances are defined by the correspondingly conditioned dependent variables. So the simulation time directly jumps from the particular time instant of a certain event to the occurrence time of the next event. **S**: event-oriented simulation

**discrete-event simulation language** A simulation language, which equalises the simulation time with the actual time of the discrete-event occurrence, considering also the criterium for the next discrete-event occurrence. **S**: discrete simulation language (2)

**discrete-event system** An event-driven system, the state transitions of which are initiated by events that occur at asynchronous discrete time instants, e.g., telecommunication network, computer network, traffic control, flexible manufacturing system, assembly system.

**discrete Fourier transform** A linear operator that converts a finite sequence of equidistant samples of a time-domain function into a same-length sequence of equidistant samples of a function with complex values in the frequency domain. **S**: DFT

**discrete process** Successive changes in an arbitrary system, where the states are unambiguously separated from each other. For every state the previous or subsequent state, or both of them, can be specified, e.g., a manufacturing line where the products are transformed, assembled, transported and stored, while their identity is not changed.

**discrete sensor** A sensor, the output of which has a finite number of states, e.g., proximity sensor, limit switch, thermostat, reed relay. Its output is often binary. **S:** on-off sensor

**discrete simulation language 1**. A simulation language for the simulation of a system, which is described by a discrete-time model. **2**. → discrete-event simulation language

**discrete-time control** A control where the control signal is sampled. **S:** sampled-data control, sampled-time control

**discrete-time controller** A controller that calculates its output considering the past sampled values of its input, usually the previous and the actual sampled value. Its output is applied in the next time step and the process is repeated consecutively. **S:** discrete controller, sampling controller

**discrete-time model** A mathematical model with dependent variables, which are defined only in the distinct equidistant time instants. It is often described with difference equations. **S:** sampled-data model

**discrete-time signal** A signal consisting of a sequence of values that are defined at discrete instants. It can be obtained by sampling a continuous-time signal, usually at equidistant time instants.

**discrete-time simulation** A simulation of a dynamic model, the states of which are changing periodically in equidistant time instants. **S:** interval-oriented simulation

**discrete-time system** A system, which operates on a discrete-time signal that causes a discrete-time signal output. It is described by a discrete-time model.

**discretisation error** An error, caused by the representation of a continuous function in a digital computer by a finite number of possible function values. It can be decreased by increasing the number of possible function values. Consequently, computational time increases.

**displacement gauge** → displacement sensor (1, 2)

**displacement sensor 1**. A sensor, which measures linear or circular motion in a certain direction, e.g., capacitive displacement sensor, inductive displacement sensor, optical displacement sensor. It is used for a wide range of purposes in industry, e.g., machine tool position control, industrial robot control, vehicle guidance and control, but also in measurements of other physical quantities that can be converted

62                                                                                          5   D

to movement, e.g., acceleration, pressure, level, temperature. **S**: displacement gauge **2**. A sensor, which measures the distance to an object. It is used in various applications where measurements of thickness, height, flatness or position are needed. **S**: displacement gauge

**displacer level sensor** A level sensor, which measures the buoyancy force of an object of constant cylindric shape partially immersed into a liquid. The buoyancy force is measured either using a torque tube with strain gauges or using a range spring with an LVDT. The changes in the buoyancy force are proportional to the measured liquid level changes. It can be used in, e.g., condensate drum, separator, storage vessels. **S**: displacer level transmitter

**displacer level transmitter** → displacer level sensor

**distal direction** A direction away from the robot base toward the robot end-effector.

**distance meter** → distance sensor

**distance sensor** A noncontact sensor, which measures the time a sound signal or a light signal needs to go from one object to another object, or from one object to another object and back, or measures the intensity of the signal, e.g., ultrasonic distance sensor, laser distance sensor. The measurement enables the calculation of the length of the signal path. It is used in, e.g., industrial applications, medical applications, military applications, machine vision, as well as in, e.g., robot, smart car, drone. **S**: distance meter, rangefinder

**distributed control system** A control system where controllers or PLCs control subsystems or groups of subsystems of a plant, usually with a large number of control loops. The control components are connected through the corresponding communication network.

**distributed-parameter model** A mathematical model that takes into account the spatial dimensions of the system and the spatial distribution of its elements, e.g., a model represented with partial differential equations with spatial coordinates and time as independent variables.

**distributed-parameter system** A system, the states of which depend on time and location. Therefore, it takes time for excitation to spread to the whole system. The system has several independent variables, usually time and one, two or three spatial coordinates. It can be modelled with partial differential equations.

**disturbance** An exogenous quantity, which enters a control loop from the outside and cannot be influenced directly, e.g., load variation, noise. It affects the controlled variable and impairs the intended control actions resulting in an increased error. It can enter the control loop at the input, at the output or anywhere in between.

**disturbance rejection** Operation of a control system, which is primarily designed to reduce the impact of any disturbance or load change on the controlled variable. On the other hand, the setpoint is expected to remain constant or not to change very often.

5 D 63

**disturbance signal** The signal that affects system operation, but cannot be influenced directly. Its consequence can be observed in the system output.

**diverter valve** A three-way valve, which can split the fluid flow from an inlet to two outlets conveying the fluid from one to two vessels. It is useful in the flow control in, e.g., chemical processing, food processing, waste processing, water delivery, medical service, pharmaceutical service. **S**: diverting valve

**diverting valve** → diverter valve

**DLMS/COSEM** ↔ device language message specification/companion specification for energy metering

**DNA method** ↔ direct-Nyquist-array method

**DOF** ↔ degree of freedom (1, 2, 3)

**domain expert** A person with special knowledge or skills in the particular field of the modelled system, who cooperates with the corresponding expert in the fields of modelling and simulation in the process of model development and model validation. This results in shortened modelling procedure as well as in a better model. **S**: application-area expert, subject-matter expert

**domestic robotics** A subfield of service robotics, in which robots perform indoor or outdoor household tasks, e.g., floor cleaning, window cleaning, lawn mowing, pool cleaning, gutter cleaning. It also encompasses toys and telepresence in a household.

**dominant pole** A pole of a transfer function, which is the closest to the imaginary axis of the $s$-plane or closest to the unit circle in the $z$-plane, hence affecting the transient response of a stable linear system the most. The slowest part of a system dominates the response, while faster parts have a minor influence on the dynamics of the system. If it is a single real pole the system behaves like a first-order system, if it is a pair of complex conjugate poles the system behaves like a second-order system.

**Doppler ultrasonic flow meter** An ultrasonic flow meter, which transmits an acoustic signal of a specific frequency into the flowstream of liquid containing suspended particles or bubbles and calculates the flow rate from the frequency shift of the reflected acoustic signal. Clamp-on construction allows quick and noninvasive installation as well as accurate measurements of diverse liquids under a variety of temperature and flow conditions.

**double-seat valve** **1**. A valve, which is constructed with two valve discs that are simultaneously moved by the valve stem. In the case of high operating pressure, it reduces the forces acting on both valve discs and on the valve stem. It eases the control valve operation, reduces the pressure drop and alleviates cavitation. **S**: two-seat valve **2**. → mixproof valve

**doublet impulse** A test signal composed of two consequent, equally sized pulses, with a positive and a negative amplitude, respectively.

**drift 1.** A gradually increasing or decreasing output signal of a sensor, which is independent of the measurement property, e.g., zero drift, operating point drift, static characteristic drift. Therefore, multiple measurements of the unchanged input signal must be performed under the same conditions. It may be caused by some environmental factors, e.g., electric field, magnetic field, thermal changes, mechanical vibrations. **2.** A property of a measuring system, expressed with a constant speed, with which the output changes at a constant, often zero value of the measured variable.

**drone 1.** A multirotor unmanned aerial vehicle, the control of which is enabled by the corresponding speed changes of the individual rotors, e.g., quadcopter, hexacopter, octocopter. **2.** → unmanned air-vehicle

**drum-type gas flow meter** → wet-test gas flow meter

**DSP** ↔ digital signal processor

**D term** ↔ derivative term

**duty cycle 1.** The time interval, during which the observed device or system is active, used or controlled by an operator. It is often expressed as a percentage. **S**: duty factor, duty ratio, power cycle **2.** The fraction of time in a period of a PWM signal, during which the signal is in the high state. It is often expressed as a percentage.

**duty factor** → duty cycle (1)

**duty ratio** → duty cycle (1)

**dyadic controller** A controller for a linear MIMO system that is described by a unity-rank matrix obtained from the product of the column vector and the row vector. Consequently, it has a specific MIMO structure.

**dynamical model** → dynamic model (1, 2)

**dynamical system** → dynamic system

**dynamic biped walking** A walking pattern, in which vertical projection of COM is not equal to ZMP and can fall outside the support polygon during some period of motion.

**dynamic error coefficient** The coefficient used for the calculation of steady-state error in a closed-loop system with the reference signal that is not described with a polynomial. **S**: generalised error coefficient

**dynamic model 1.** A mathematical model of a dynamic system, which describes the system behaviour in the transient response and in the steady state with, e.g., differential equations, difference equations. **S**: dynamical model **2.** A physical model, which can change its internal states with regard to time, e. g., prototype, pilot plant, laboratory setup. **S**: dynamical model

**dynamic movement primitives** A framework for learning point-to-point trajectory from a demonstration. It adapts the trajectories measured on a human, subject to the kinematic and dynamic capabilities of a humanoid robot. The approach is based on solutions of nonlinear differential equations that describe smooth kinematic robot-joint trajectories.

**dynamics** Time-dependent system behaviour, described by, e. g., a differential equation, a difference equation, a time response, a frequency response.

**dynamic system** A system, in which time-dependent qualitative or quantitative changes are taking place. The output of such a system depends not only on the current input value but also on the previous input values. **S**: dynamical system

**dynamic variable** An element of a stock and flow diagram denoting a value that influences other elements. It can be either a function of stocks, a constant, or an exogenous input. It is used as a source or an intermediate converter of information and is usually depicted as a circle. **S**: converter (2)

# Chapter 6
# E

**eddy current displacement sensor** → inductive displacement sensor (2)

**eddy current proximity sensor** → inductive proximity sensor

**educational robotics** A subfield of robotics that deals with the teaching of analysis, design, application and operation of robots, mobile robots or autonomous vehicles. The corresponding devices consist of hardware, e.g., preassembled device, kit of components, and software, containing source code and programming environments. It can be taught from elementary school to graduate programs and is often applied in robot tournaments.

**effective workspace** → robot workspace

**efflux viscometer** A viscometer, which measures the time needed to drain a specified volume of fluid through a fixed orifice at the bottom of a vessel, e.g., Saybolt viscometer, orifice viscometer, redwood viscometer. The time is proportional to the measured viscosity at the controlled temperature. It is used in measurements of, e.g., oil, syrup, varnish, paint, bitumen emulsion.

**eigenfrequency** → natural frequency

**eigenvalue assignment 1.** → full-state feedback **2.** → pole assignment (2)

**elasticity** → flexibility (1)

**elbow flow meter** A simple differential-pressure flow meter, in which a centrifugal force of a fluid passing through a curved pipe is exerted along the outer edges of the curve. Therefore the pressure at the outer wall is higher than the pressure at the inner wall. The differential pressure, measured at the curvature of the pipe, enables the calculation of the measured flow rate. The existing curves in piping systems can also be exploited.

© ZRC SAZU/Research Centre of the Slovenian Academy of Sciences and Arts 2023
R. Karba et al., *Terminological Dictionary of Automatic Control, Systems and Robotics*,
Intelligent Systems, Control and Automation: Science and Engineering 104,
https://doi.org/10.1007/978-3-031-35755-8_6

**electrical conductivity meter** → electrical conductivity sensor (1, 2)

**electrical conductivity probe** → electrical conductivity sensor (1, 2)

**electrical conductivity sensor** **1**. A sensor, which measures the ability of electric current to flow in the measured solution. It consists of two noble-metallic electrodes that are submerged in the measured solution. The applied known voltage and the measured current enable the calculation of the solution's resistance using Ohm's law, considering the known distance between the electrodes and their total immersed surface. Corresponding temperature compensation is needed. **S**: electrical conductivity meter, electrical conductivity probe **2**. A sensor, which measures the ability of electric current to flow in the measured solution. The noncontacting inductive sensor utilises the induced current in a structure of two toroidal coils that is directly proportional to the resistance of the solution. It is used in, e.g., industrial applications, aquaculture, agriculture, environmental monitoring, wastewater monitoring. **S**: electrical conductivity meter, electrical conductivity probe

**electrical grid** → power grid

**electric motor** An electromechanical energy converter, which converts electrical energy into circular or linear motion, e.g., DC motor, AC motor, stepper motor. It is used in, e.g., fan, pump, machine tool, disk drive, electrical watch, and as an actuator or a final control element in an actuator system.

**electro-hydraulic converter** → electro-hydraulic servo valve

**electro-hydraulic servo valve** An interface between electronic and hydraulic components in a system, which receives an electrical input, often from the output of the controller, converting it to a proportional hydraulic output, which is connected to, e.g., hydraulic actuator input. It enables smooth, high-strength motion in a hydraulic system. It is used in, e.g., power plant, furnace, cargo crane, mobile vehicle, flight simulator. **S**: electro-hydraulic converter, electro-hydraulic transducer

**electro-hydraulic transducer** → electro-hydraulic servo valve

**electromagnetic flow meter** An obstruction-less flow meter, which measures volumetric flow rate and operates under the induction principle. It consists of a nonconductive tube or conductive tube with an insulated lining and outer coils creating a magnetic field. The latter is perpendicular to the conductive fluid flow and the axis of the electrodes. The induced voltage is proportional to the average fluid velocity and consequently also to the measured volumetric flow rate. It measures the flow rate of, e.g., sanitary liquids, dirty liquids, corrosive liquids, abrasive liquids, slurries. **S**: magmeter, magnetic flow meter

**electromechanical energy converter** A device, which enables the interchange of energy between mechanical and electrical systems, and vice versa. Mechanical energy is converted into electrical energy by, e.g., a generator, a microphone, a piezoelectric transducer. Electrical energy is converted into mechanical energy by, e.g., an electric motor, a solenoid, a speaker.

**electronically commutated motor** → brushless DC motor

**electro-pneumatic converter** → electro-pneumatic transducer

**electro-pneumatic transducer** An interface between electronic and pneumatic components in a system, which receives an electrical input, often from the output of the controller, converting it to a proportional pneumatic output, which is connected to, e.g., pneumatic actuator input. It is used in, e.g., petrochemical industry, paper industry, food industry, energy management, heating system, air-conditioning system, medical system. **S**: electro-pneumatic converter

**element** **1**. The indivisible part of the hardware in a control system. **2**. The part of a system that can be implemented to fulfill specified requirements, usually contributing to the function of the whole system. **3**. An object or entity, dealing only with inputs and outputs. Its internal mechanisms are not considered.

**element zero** The zero of an individual transfer function of the TFM of a MIMO system. Its role in the MIMO-systems theory is the same as in the SISO-systems theory.

**elevation** → pitch angle

**embeddable technology** → implantable technology

**embedded system** A computer system, the core of which is a processor, e.g., micro-controller, DSP, application-specific integrated circuit, PLD. Furthermore, it includes memory and peripheral devices to handle various electrical and mechanical interfaces. It is aimed only for a small number of functions within a larger mechanical or electrical system. Its size, power consumption, performance, reliability and cost can be optimised according to the requirements of the host system. It is often used in real-time operation in, e.g., control system, medical equipment, hybrid vehicle, consumer electronics, telecommunications, avionics.

**emergency power-off** → emergency stop

**emergency stop** A safety mechanism, which completely aborts the operation of a device or machinery in unpredictable or dangerous circumstances. Shut down must be as quick as possible but at the same time easily realisable even for an untrained operator or bystander. **S**: emergency power-off, kill switch

**empirical validity** A model-validation procedure, which enables the quantitative determination of the degree of fitness between model response and measured data or prescribed curve. For this purpose statistical methods, e.g., analysis of variance, factor analysis, spectral analysis, regression, are used. **S**: validity of results

**encoder** An electro-mechanical sensor, which converts a position or an orientation of the object into a corresponding electrical signal, e.g., absolute encoder, incremental encoder, linear encoder, rotary encoder. It is frequently used in the process industry and robotics.

70                                                                                          6  E

**end-switch**  → limit switch

**energy balance** The conservation law, which states that time derivative of stored energy in an energy storage system is equal to the difference between the sum of energy flows at the input and the sum of energy flows at the output of the energy storage system. **S**: law of conservation of energy

**energy density spectrum**  → energy spectral density

**energy-dissipation element** An element, which causes damping in a system by transforming energy from one form to another form, e.g., damper, resistor, valve. **S**: energy dissipator

**energy dissipator**  → energy-dissipation element

**energy source**  → power source (1, 2)

**energy spectral density** A function decribing a frequency-dependent representation of signal energy. The integral of the function along an arbitrary frequency interval is a constituent part of the whole energy of the signal. It is suitable for signals with finite energy, e.g., transient response. **S**: energy density spectrum

**engineering unit** A unit, which does not belong to base units or derived units, e.g., tonne, litre, calorie, atmosphere, minute, bar, degree Celsius, foot, pound, pint. It is often used in industrial practice. **S**: non-SI unit

**enterprise resource planning** A modular software designed to manage all information and resources involved in the operation of a company. It integrates planning, purchasing inventory, sales, marketing, finance, human resources, processes and technologies, across a modern enterprise into a unified system. It enables higher productivity, improved agility, better insight, enhanced collaboration and lower risk. **S**: ERP

**equal-percentage valve characteristic** An inherent valve characteristic, which shows an exponential relation between the flow rate and the valve travel. Any percentage change in the valve travel from its current value changes the flow rate by the same percentage of its current value, regardless of the position in the operating range. It is used in valves for temperature or pressure control, especially when large changes of pressure drop across the valve are expected.

**equation error** The difference between model predictions and observations obtained from the modelled system. It depends on input-signal value and output-signal value. **S**: input-output error

**equation-oriented simulation language** A continuous simulation language, which enables the simulation of models that are given in the form of an appropriate mathematical structure. It is supported by the syntax of the target language of the compiler. The ability of direct inclusion of differential equations in a program results in a short, well understandable and modular programs.

**equilibrium point** A point in the state space, at which a dynamic system stays if it starts from it.

**equilibrium state** A state of a system, which does not change, if there are no external excitations. It is stable when a system always returns to it after small perturbations, and it is unstable when a system moves away from it after small perturbations.

**ergonomics** Scientific discipline, which studies interactions among humans and particular components of the system, applying psychological and physiological principles to the process of product design or system design. It tends to assure a convenient working environment for the operator, easy and safe handling of the system as well as the optimal operation of the system.

**ERP** ↔ enterprise resource planning

**error** 1. The difference between the setpoint value and the controlled-variable value in a control system in the case of unity feedback. 2. The difference between the setpoint value and the output value of the signal modifier or controller in the feedback path. 3. The difference between the computed, estimated or measured value, and the defined, prescribed or theoretically correct value of a variable or of a parameter.

**estimator** A system that is used for calculating the estimated value of the considered quantity from measurements or observations.

**EtherCAT** ↔ Ethernet for control automation technology

**Ethernet** A group of networking technologies for connecting computers and devices, operating on the physical layer and data-link layer. It enables the creation of the most commonly used wired, high-speed, reliable and secure local area networks, which use connections with coaxial cables, twisted-pair cables or optical fibres. The physical distance limit can be extended using routers, switches and hubs.

**Ethernet for control automation technology** A master-slave industrial Ethernet for real-time high-speed data communication in automation. It is an open fieldbus-based technology and offers low-cost implementation, a high level of synchronisation accuracy, comfortable system diagnostics and easy deployment. **S**: EtherCAT

**Ethernet industrial protocol** An industrial Ethernet for transferring large amount of data at the application layer of Ethernet combined with TCP. It merges the advantages of the internet and open technologies. It is easy to configure, operate, maintain and scale-up. Therefore, it is used in a wide range of control applications and information data exchanges as well as in IIoT. **S**: EtherNet/IP

**EtherNet/IP** ↔ Ethernet industrial protocol

**Euler angles** A set of three angles, which fully determine the orientation of an object in space taking into account different conventions, depending on the order of rotations.

**Euler method** A first-order numerical integration method for solving ordinary differential equations, which uses a basic rectangular single-step approximation of the integral. For the given initial value and step size, the differentials in derivatives of the differential equation are replaced with differences.

**event-oriented simulation** → discrete-event simulation

**event-oriented simulation language** → discrete-event simulation language

**evolutionary algorithm** A stochastic searching method, which imitates natural biological evolution, e.g., genetic algorithm, PSO. This heuristic-based approach contains phases of, e.g., initialisation, selection, genetic-operators acting and termination. It comprises a variety of algorithms providing an alternative to the conventional way of solving searching problems.

**evolving model** A mathematical model, which improves the fit of its response to measured data by changing both its structure and its parameters using online identification. Its structure and parameter values are adjusted correspondingly as new data become available.

**exhaustive search** → brute force

**experiment** The process of obtaining data from the system excited through its input. The data are required in, e.g., mathematical-model development and its validation, control system design and its validation.

**experimental modelling** → system identification

**expert system** A computer program that emulates the analysing, solving and decision-making abilities of a human expert for a particular problem domain. It incorporates a knowledge base and an inference engine or rules engine, which enables applying the knowledge base to each particular problem. So the wide use of built-in knowledge is possible, which facilitates the work of less experienced users as well as of experts.

**explicit enumeration** → brute force

**extended Kalman filter** A Kalman filter for the calculation of optimal estimates of the unknown variables, frequently system states, intended for nonlinear dynamic systems. It employs the linearisation of the system in the current mean estimate and in the covariance estimate.

**external sensor** A sensor, which provides information about the environment of a control system or a robot manipulator. It is not physically attached to the control system nor to the robot manipulator itself. It can be, e.g., digital camera, range sensor, contact sensor, proximity sensor.

**exteroception** **1**. Sensitivity to stimuli originating outside of the body. **2**. Acquisition of the information about the environment of a robot using the sensors that are not mounted on the robot itself.

**extrapolation integration method** A numerical integration method, which uses different calculation subintervals where every subinterval is smaller than the preceding one, e.g., Richardson extrapolation, rational extrapolation, Euler-Romberg method, Bulirsch-Stoer-Gragg method. Multiple low-accuracy integrations on the calculation subintervals enable highly accurate results.

# Chapter 7
# F

**factored transfer-function form** → zero-pole-gain transfer-function form

**factor-label method** → dimensional analysis (1, 2)

**falling-ball viscometer** A viscometer, which measures the time required for a solid ball to descend through a sample-filled tube, affected only by gravitational, buoyant and frictional forces. The known diameter and density of the solid ball as well as the diameter and length of the tube enable the calculation of the measured viscosity. **S**: falling-sphere viscometer

**falling-piston viscometer** A viscometer, which measures the time required for a piston to descend in a sample-filled cylinder, affected only by gravitational, buoyant and frictional forces. A lifting mechanism raises the piston and draws the tested medium into the cylinder. The time of the descent of the piston enables the calculation of the measured viscosity.

**falling-sphere viscometer** → falling-ball viscometer

**fast Fourier transform** An algorithm for the calculation of a DFT or its inverse. It reduces the number of necessary operations and consequently reduces the complexity of the calculation as well as increases the efficiency of the algorithm's implementation. **S**: fast Fourier transformation

**fast Fourier transformation** → fast Fourier transform

**fast opening characteristic** → quick-opening characteristic

**fault detection** The analysis, which reveals whether a fault has occurred in a system, e.g., controlled process, software. The fault occurrence time is also estimated.

**fault diagnosis** The determination of the class, place, size and dynamics of a physical fault. It comprises fault isolation and fault identification, and is carried out after fault detection.

© ZRC SAZU/Research Centre of the Slovenian Academy of Sciences and Arts 2023
R. Karba et al., *Terminological Dictionary of Automatic Control, Systems and Robotics*,
Intelligent Systems, Control and Automation: Science and Engineering 104,
https://doi.org/10.1007/978-3-031-35755-8_7

**fault identification** The estimation of the size and dynamics of a fault. It follows the procedure for fault isolation.

**fault isolation** The identification of fault class and the place of fault occurrence. It is carried out after fault detection.

**fault-tolerant control** A set of methods that are tolerant to the occurrence of simple faults in a closed-loop system and prevent that simple faults result in a serious failure. It comprises fault detection and possibly fault diagnosis, as well as suitable control-design methods.

**feasibility study** The analysis evaluating a realisability of project, which takes into account all relevant factors.

**feedback compensation** $\rightarrow$ parallel compensation

**feedback control** A control strategy, which compares the actual value of the controlled variable with its desired value. Based on the obtained error, the control-variable value is calculated, which tends to reduce the error to 0. **S**: closed-loop control

**feedback loop** A closed path between two subsystems, which affect each other, e.g., control loop, biological feedback loop, dialogue.

**feedback path** A path in a graphical representation of a system, e.g., in a signal-flow graph, in a block diagram, which is routed back from the output or some other node to the input. There can be more than one in a system. **S**: reverse path

**feedforward** A path within a signal-flow graph or within a block diagram that is parallel to the main forward path. It is often the signal path from the setpoint to the control input.

**feedforward control** An open-loop control, in which the disturbance measurement enables a corresponding corrective action. The corrective action is taken before the disturbance could affect the behaviour of the system, therefore effectively compensating for the disturbance. In practical applications, it is most often combined with feedback control. In such configuration, it reduces the effect of measurable disturbances, while the feedback part compensates for measurement errors, for unmeasurable disturbances or sometimes for inaccuracies in the process model.

**feedforward matrix** The matrix in the output equation of a state-space model, which defines the relationship between the system inputs and the system outputs. It is usually marked with the letter $D$. Its elements are frequently equal to 0. For a SISO system, it is reduced to a scalar. **S**: direct matrix, feedthrough matrix

**feedforward neural network** An ANN, the elements of which are organised in layers and connected in the way that data flow only in the direction from input to output with no loops, e.g., multilayer perceptron, probabilistic neural network, radial basis-function network.

**feedforward transfer function** A transfer function, which dynamically relates the controlled signal to the error signal.

**feedthrough matrix** → feedforward matrix

**FEM** ↔ finite element method

**fembot** → gynoid

**fibre-optic displacement sensor** A noncontact displacement sensor, which transmits a beam of light through a flexible optical probe, receives the light reflected from the target using the corresponding photosensor and converts this light into an electrical signal proportional to the distance between the probe and the measured target. It is small, reliable, accurate, sensitive and usable in, e.g., automotive applications, robotic applications, aeronautic applications, aerospace applications, high-voltage applications, nuclear applications.

**fieldbus** A group of industrial-network protocols, used for real-time digital transmission in distributed manufacturing and process control, e.g., Profibus, CC-link, Modbus, HART protocol, EtherCAT. Sensors, actuators and controllers can be connected in different network topologies, e.g., ring, branch, star, daisy chain.

**field-programmable gate array** A PLD, which operates as a matrix of logic blocks that also include memory elements and are connected via programmable interconnections, thus enabling the generation of complex logic functions. It is used in, e.g., industrial systems, embedded systems, automotive systems, image processing, wireless communications, defence, bioinformatics. **S**: FPGA

**filled system thermometer** → pressure spring thermometer

**final control element** The part of an actuator system, which acts directly on the controlled system, e.g., valve, flap, heater, pump, motor. It must be compatible with the actuator output and with the controlled system input causing changes in the corresponding quantities and consequently in the controlled variable.

**final control system** → actuator system

**finite element analysis** → finite element method

**finite element method** A simulation method, which divides the mathematical model that represents a component or an assembly into small sub-areas. The behaviour of each particular sub-area is simulated individually, thus yielding the results, which are merged to encompass the whole component or assembly. It is used in, e.g., fluid dynamics, structural analysis, thermal transport. **S**: FEM, finite element analysis

**finite impulse-response model** A regression model, which represents the mapping of the present output value from a finite number of previous values of inputs and from the present value of the input. Its coefficients are optimised according to a one-step-ahead prediction error. **S**: FIR model

**FIR model** ↔ finite impulse-response model

**first controller** → master controller

**first method of Lyapunov** → Lyapunov's indirect method

**first-order hold** An element that converts a discrete-time signal to a continuous-time signal by linearly interpolating between its sampled value in the actual sampling instant and its sampled value in the next sampling instant. As the sampled value in the next sampling instant is not available online, it is an acausal conversion. **S**: FOH

**first-order system** A dynamic system, which is described with a first-order differential or difference equation.

**first-principles modelling** Methodology for building a mathematical model of a dynamic system, which is based on partitioning of the modelled system into appropriate subsystems. The latter and their interconnections are determined with the domain laws, e.g., phenomenological laws, conservation laws. Every relevant subsystem is described by the coresponding mathematical relations. Consequently, the final result is a system of equations. **S**: analytical modelling

**Fisher information matrix** A matrix describing the amount of information about an unknown parameter, which is carried by an observable random variable. It is inversely proportional to the variance of the observed parameter or its expected value.

**fitness function** An objective function, the bigger value of which represents a better result. Therefore, in an optimisation problem, its value is maximised. **S**: profit function, reward function, utility function

**fixed PLC** → compact PLC

**fixed-step integration method** A numerical integration method with a constant step size during the whole simulation run, e.g., single-step method, multi-step integration method.

**fixture** A mechanical device, which firmly constrains a workpiece in the desired pose.

**flapper-nozzle amplifier** A pneumatic or hydraulic amplifier, which consists of a compressed-fluid supply connected to a nozzle. A small displacement of a flat plate mounted in front of the nozzle causes a corresponding output-pressure gain. It is often used in a valve positioner. Moreover, it is used for converting pressure to mechanical motion and vice versa or for converting current to pressure when the plate position depends on the current signal.

**flap valve** A valve, which allows the flow of fluid in one direction, while preventing the flow in the opposite direction. A plate, hinged on one edge and mounted across a tube or a duct, is either forced open by the flow or opened remotely. It is used in, e.g., water-treatment plant, irrigation system, industrial-waste line, flood control.

**flexibility 1**. The property of an object, reciprocal of stiffness, which causes an elastic body to return to its original shape after the influence of force or torque ceases. **S**: compliance (2), elasticity **2**. The ability of a robot to perform several kinds of different tasks. **3**. The ability of the robot arm to bend in almost any direction. **4**. The property of an industrial robot that is elastic at its joints or along its links.

**flexible design and manufacturing** $\rightarrow$ computer-integrated manufacturing

**flicker noise** $\rightarrow$ pink noise

**float level sensor** A level sensor, which is used for continuous measurements or as a switch, detecting measured level from the position of a corresponding buoyant object, e.g., ball float level sensor, chain float level sensor, magnetic float level sensor. It is used in commercial and industrial applications for various liquids.

**flow** An element of a stock and flow diagram that influences the rate of change of the value of the stock, to which it is connected. It is usually depicted as a thick arrow with a valve. If the arrow points into a stock, it causes the value of the stock to increase, if it points out from a stock, it causes the value of the stock to decrease. **S**: rate

**flowchart** A diagram that graphically shows the activities and decisions involved in a process, algorithm, model or computer program. The particular steps are represented by various shapes that are interconnected with arrows denoting the direction of the execution. **S**: flow diagram

**flow coefficient** The parameter of a valve, which is used to determine the proper valve size for a given set of service conditions. It defines the volumetric flow rate in gal/min of water with a prescribed temperature, density and dynamic viscosity at 1 PSI pressure drop. It can be defined either for a fully open valve as well as for a specific valve-stem position. When dealing with a fluid other than water, its specific density and dynamic viscosity have to be taken into account. If required, it can be converted to a flow factor. **S**: valve sizing coefficient

**flow diagram** $\rightarrow$ flowchart

**flow factor** The parameter of a valve, which is used to determine the proper valve size for a given set of service conditions. It defines the volumetric flow rate in $m^3$/ h of water with a prescribed temperature, density and dynamic viscosity at 1 bar pressure drop. It can be defined either for a fully open valve as well as for a specific valve-stem position. When dealing with a fluid other than water, its specific density and dynamic viscosity have to be taken into account.

**flow meter** A sensor, which measures volumetric flow rate or mass flow rate of fluids, e.g., positive-displacement flow meter, electromagnetic flow meter, ultrasonic flow meter, turbine flow meter, rotameter, vortex meter, Coriolis mass flow meter. It is one of the most commonly used measuring instruments in industry. **S**: flow sensor

**flow nozzle** A differential-pressure flow meter with a specifically-constructed restriction, which is mounted between two flanges or is welded inside the pipe. The fluid enters smoothly and comes to the cylindrical throat where the cross-section area is minimal. Its properties are between the properties of the orifice-plate flow meter and the Venturi flow meter.

**flow of information** → link

**flow sensor** → flow meter

**flow switch** A device, which senses the flow of fluid and changes its binary output at the moment of flow appearance or disappearance. It is commonly used in general industrial applications as well as in, e.g., pump protection, heat exchanger, cooling water system, safety valve monitoring.

**fluid** A substance, which has no fixed shape. It is capable of flowing and tends to assume the shape of its container, e.g., liquid, gas, steam, plasma, slurry, granulate.

**fluorometer** An extremely sensitive photometer, which measures parameters of visible spectrum fluorescence, e.g., filter fluorometer, spectral fluorometer. Intensity and wavelength distribution of emission spectrum after excitation by a specific spectrum of the ultraviolet light or mercury-lamp light enables the determination of the presence and the amount of certain responsive molecules in a sample. The concentrations of unknowns can be mathematically quantified against the fluorescence intensity of known molecules in the sample. It is used in, e.g., bioengineering, pharmaceutical QC, environmental monitoring, medicine.

**flywheel** An idealised lumped-parameter element for modelling a rotational mechanical system, which stores rotational energy and has the property of inertia, e.g., RW. **S**: angular mass

**$1/f$ noise** → pink noise

**focus point** An equilibrium point of a second-order system in the phase plane, which is a source or a sink of trajectories in the spiral form. **S**: spiral point

**FOH** ↔ first-order hold

**FOH equivalent** A time-response fitting method for the system discretisation, which results in a discrete-time model that has the same response in the sampling instants as the original continuous-time model for piecewise linear inputs.

**force balance** The momentum balance described with the second Newton's law. The time derivative of momentum expressed as the product of mass and acceleration is equal to the sum of external forces.

**force closure** The ability of the robot grasp to inhibit motion of the workpiece using contact friction despite arbitrary externally applied forces.

**force control** Control of the robot, which reduces the difference between the desired force and the measured force at the robot end-effector. It is mandatory for achieving robust and versatile behaviour of a robot system in poorly structured environments as well as a safe and reliable operation in the presence of humans.

**force – current analogy** An analogy between a translational mechanical system and an electrical system, which results in additional analogue pairs, i.e., mass – capacitance, damper – reciprocal of resistance, spring – reciprocal of inductance, displacement – magnetic flux, linear velocity – voltage.

**forced response** A time response of a dynamic system, which is excited solely with the external input. The initial conditions equal 0. It represents a particular solution of the differential equation that describes the mathematical model of the observed system.

**force reflection** A teleoperation, in which the operator feels the force exerted by the telemanipulator to the environment using a haptic interface.

**force-torque sensor** A sensor in a robot wrist, which measures the external physical interaction between robot end-effector and its environment in three orthogonal directions. Forces are often measured with strain gauges. It is often applied in collaborative robots.

**force – voltage analogy** An analogy between a translational mechanical system and an electrical system, which results in additional analogue pairs, i. e., mass – inductance, damper – resistance, spring – reciprocal of capacitance, displacement – charge, linear velocity – current.

**forgetting factor** A parameter in online data processing, which defines the decrease rate of the influence of the older incoming data on the actual result.

**form closure 1**. A geometric property of robot grasp, described by a complete constraint of the grasped object. **2**. The ability of the robot grasp to prevent motions of the workpiece by frictionless contact.

**forward channel** → forward path

**forward dynamic model** → direct dynamic model

**forward kinematics** → direct kinematics

**forward path** A path in the graphical representation of a system, e.g., in a signal-flow graph, in a block diagram, which connects the input to the output without passing any single node or path more than once. There can be more than one in a system. **S:** forward channel

**forward rectangular rule** A frequency-response fitting method for the system discretisation, which approximates the derivative of a variable with the difference between its value in the next time step and its value in the actual time step, divided by sampling time.

**forward selection** An iterative method for the selection of input variables of a mathematical model of a system, where the variables are systematically added to the model. The addition of each variable is tested with a chosen model-fit criterion.

**FPGA** $\leftrightarrow$ field-programmable gate array

**free response** $\rightarrow$ natural response

**frequency characteristics** $\rightarrow$ frequency response

**frequency domain** Analytic space, in which the set of values that are accepted as the function input is covered by frequency as the independent variable.

**frequency response** The steady-state response of a system to sinusoidal input signals of selected frequencies. It can be either measured, obtained by simulation or calculated analytically. It is often represented graphically, e.g., with a Bode plot, with a polar plot. **S**: frequency characteristics, harmonic response

**frequency-response fitting** A method for the system discretisation, which results in a discrete-time model with similar frequency-related properties as the original continuous-time model, e.g., forward rectangular rule, backward rectangular rule, bilinear transformation.

**frequency spectrum** $\rightarrow$ spectral density

**frequency warping** A distortion in the frequency response of a discrete-time model, which results in the same frequency response as the frequency response of the equivalent continuous-time model at a higher frequency. It is more notable in higher frequencies close to Nyquist frequency.

**friction** A force, which resists the relative motion or the tendency of two objects to slide against each other, e.g., dry friction, static friction, viscous friction. It converts kinetic energy to thermal energy in, e.g., damper, rotary damper.

**friction cone** A graphical representation of the Coulomb-friction model. The set of all external forces that can be applied to the object by the point contact without the object slipping is constrained inside it.

**Frobenius canonical form** $\rightarrow$ companion form

**Frobenius normal form** $\rightarrow$ companion form

**fruitfulness** **1**. The property of a model, that the conclusions obtained from analyses of model results are useful. **2**. The property of a model, that the conclusions obtained from analyses of model results motivate the development of other good models.

**fuel cell** An electrochemical unit, which converts the chemical energy of a fuel and oxidising agent into electricity through a pair of redox reactions, e.g., alkaline fuel cell, phosphoric-acid fuel cell, molten-carbonate fuel cell, proton-exchange-membrane fuel cell. It is often used as a power supply of an actuator system.

**full-order observer** A system for the estimation of the values of all system states, both nonmeasured and measured. It gathers information from measured signals, usually the input signal and the output signal.

**full-state feedback** A family of control-design methods for computing parameters of the SISO-system controller according to the requirements concerning the location of poles of the closed-loop system in the $s$-plane. The resulting controller in the negative feedback loop generates the control signal, which is obtained as a linear combination of all the state variables weighted by a coefficient matrix. **S**: eigenvalue assignment (1), pole allocation (1), pole assignment (1), pole placement (1), pole shifting (1)

**full-state feedback controller** A state controller, the input of which is the whole set of the actual states of the controlled system. **S**: pole-placement controller

**functional controllability** The property of a linear MIMO system that enables the choice of the corresponding inputs, causing the desired outputs for zero initial conditions. The necessary and sufficient condition is the nonsingularity of the TFM of the system. **S**: functional reproducibility

**functional reproducibility** → functional controllability

**function block** The single-input or multiple-input block of a simulation scheme, which represents an arbitrary input-signal source, a generator of arbitrary input-output relations, mathematical operations or mathematical functions.

**function-block diagram** A graphical standard PLC programming language that depicts the connections between logic gates and other blocks.

**function generator** **1**. Electronic test equipment or an electronic device used to generate various types of signal waveforms, e.g., sine wave, square wave, triangular wave, over a wide range of frequencies. It is used in simulation, as well as in the development, testing and repair of electronic devices. **2**. Software or a software module used to generate various shapes of signals with interpolation between predefined points in the time domain.

**fundamental physical dimension** → basic physical dimension

**fundamental quantity** → base quantity

**fundamental unit** → base unit

**fuzzification** The operation, which considers a certain variable and assigns to it the appropriate membership values with regard to each fuzzy set. The membership values are assigned according to the numerical value of the variable.

**fuzzy associative memory** → fuzzy model

84                                                                    7 F

**fuzzy control** Control using algorithms that involve fuzzy logic. It comprises a variety of algorithms ranging from linear algorithms to adaptive algorithms.

**fuzzy controller** A controller, which generates the control variable using fuzzy logic.

**fuzzy identification** System identification, commonly of a nonlinear system, which results in a fuzzy model, e.g., Takagi-Sugeno fuzzy model. The identification of the model structure includes the definition of the type and the number of rules or membership functions, while parameter estimation is used to obtain the values of the corresponding parameters.

**fuzzy inference system** → fuzzy model

**fuzzy logic** A generalisation of the classic binary logic, in which a logical variable or a logical statement can have any value from the interval between 0 and 1. It is an attempt to utilise imprecise information in mathematical formulations. Its tools enable the transformation of a linguistic description into an algorithm, the result of which is an action.

**fuzzy model** A mathematical model that consists of membership functions, fuzzy operators and fuzzy rules, which implement fuzzy logic to fully define the relations between the inputs and the outputs of the modelled system, e.g., Takagi-Sugeno fuzzy model, Mamdani fuzzy model. It is often used for modelling nonlinear dynamic systems providing explicit and understandable representation of knowledge. **S:** fuzzy associative memory, fuzzy inference system, fuzzy rule-based system

**fuzzy operator 1.** A mapping or function that acts on variables or statements in fuzzy logic, e.g., fuzzy conjunction, fuzzy disjunction, fuzzy complement. **2.** An operation that acts on fuzzy sets, e.g., fuzzy union, fuzzy intersection, fuzzy complement.

**fuzzy rule** An if-then rule, which describes relations in a particular subdomain of a fuzzy model.

**fuzzy rule-based system** → fuzzy model

**fuzzy set** A generalisation of the classic binary set, a particular element of which can be its member with a certain membership degree between 0 and 1, and not only its member or not its member.

**fuzzy system** A system, which is in part or completely described by a fuzzy model.

# Chapter 8
# G

**gain** 1. The ratio between the amplitudes of the output signal and the input signal of a system in a steady state, which in the case of sinusoidal components in the input signal depends on the frequency. **S**: amplification 2. An element of a signal-flow graph, which represents the weight of a branch. In control structures, it is often a transfer function.

**gain block** The block of a simulation scheme, the output of which is the product of the input signal and a predefined constant.

**gain crossover frequency** Frequency at which the amplitude response of a dynamic system reaches unity gain, i.e., 0 dB. **S**: unity-gain frequency

**gain margin** The distance of the amplitude response of a dynamic system to the stability margin. It is a measure of relative stability.

**gain scheduling** An adaptive control, which changes the controller parameters according to operating conditions. It results in a nonlinear controller, which comprises a set of linear controllers designed for the entire range of operating conditions. **S**: parameter scheduling

**gantry robot** A big Cartesian-like robot, which has at least three prismatic joints. It is mounted overhead, enabling the handling of heavy payloads in a large reachable workspace. It is used in, e.g., pick-and-place, welding.

**gas chromatograph** A chromatograph, which separates and analyses compounds that can be vapourised without decomposition. The analysed sample is mixed with a carrier, often helium or nitrogen, and injected in a column that is placed in a temperature-controlled oven. The mixture is pushed through the column, coated with the stationary phase, by a controlled pressure and detected with mostly flame ionization detector or thermal conductivity detector. The resulting chromatogram is a graph with a spectrum of peaks enabling the identification of components in the

© ZRC SAZU/Research Centre of the Slovenian Academy of Sciences and Arts 2023
R. Karba et al., *Terminological Dictionary of Automatic Control, Systems and Robotics*,
Intelligent Systems, Control and Automation: Science and Engineering 104,
https://doi.org/10.1007/978-3-031-35755-8_8

sample, while the areas under the peaks are proportional to the amount and in turn to concentrations of the components. It is used in, e.g., QA of products in chemical and pharmaceutical industries, measurement of toxic substances in soil, air or water, analysis of pollutants, analysis of specimens in forensic sciences.

**gas detector** $\rightarrow$ gas sensor

**gas flow meter** A flow meter, which determines the flow rate of a fluid using positive displacement, differential pressure, variable area, turbine, vortex shedding, thermal convection, ultrasound, Coriolis force, e.g., wet-test gas flow meter, bellows gas flow meter. The volume of the measured fluid, e.g., natural gas, waste gas, flue gas, is highly affected by temperature and pressure and, consequently, needs a corresponding compensation.

**gasoline generator** $\rightarrow$ petrol generator

**gas sensor** A sensor, which detects and measures the presence or concentration of volatile substances in the environment, e.g., electrochemical gas sensor, optical gas sensor, calorimetric gas sensor, ultrasonic gas sensor. It can detect, e.g., smoke, oxygen, carbon dioxide, hydrogen, carbon monoxide, ozone, methane, flammable gas, combustible gas, toxic gas. It is used in, e.g., industry, household, air conditioning, fire detecting, public security, safety at work, agriculture. **S**: gas detector

**gate valve** A valve, which changes the flow by lifting a round or rectangular wedge out of the fluid path. It is predominantly used to completely shut off the fluid flow, while in the fully-open position it represents a negligible obstruction in the flow path, thus having a very small pressure drop. It is mostly used in pipes with larger diameters, e.g., in petroleum industry, in water supply, in wastewater system. **S**: sluice valve

**gauge pressure** Pressure measured relative to atmospheric pressure. It is positive for pressure above atmospheric pressure and negative for pressure below atmospheric pressure.

**Gaussian noise** Noise that has the probability density function equal to that of a normal distribution, i.e., Gaussian distribution.

**Gaussian white noise** A white noise, which has the probability density function equal to that of a normal distribution, i.e., Gaussian distribution. In practice, the generated noise is white noise only within the frequency range of interest.

**gear flow meter** A positive-displacement flow meter, which measures volumetric flow rate, e.g., oval gear flow meter, helical gear flow meter, lobed impeller flow meter. It creates a precise volume of fluid with each revolution of mostly two impellers, e.g., toothed oval wheels, lobed impellers, cogwheels. The two impellers are driven by a flowstream of fluid pushing it to the output of a sensor, where each rotation is counted.

**gear pump** A positive displacement rotary pump, which transfers fluid using cog-wheels, e.g., internal gear pump, external gear pump. It is applicable in the cases where pulseless and stable flow, accurate dosing or high-pressure output is required. It can convey high-viscosity fluids as well. It is used in, e.g., chemical industry, agricultural industry, food industry.

**generalised error coefficient** $\rightarrow$ dynamic error coefficient

**generality** The property of a model that defines the range of problem domains for which the model is applicable.

**general-purpose interface bus** A digital, parallel, short-range hardware interface for high-speed communication between an instrument and a controller. It enables the connection of multiple instruments to a single controller, while galvanic isolation from the instruments is not required. **S**: GPIB, IEEE-488

**general-purpose simulation language** A simulation language, which is intended for the simulation of mathematical models that are relatively independent of the simulated system type. Therefore it is applicable in several engineering and nonengineering domains.

**generate and test** $\rightarrow$ brute force

**genetic algorithm** Evolutionary algorithm, which imitates the process of natural selection and bases on the phases of the initial population, fitness function, selection, crossover and mutation. It is usually used for stochastic optimisation.

**global asymptotic Lyapunov stability** The property of an autonomous system, the solutions of which start out from any initial point in the state space, stay within a limited proximity of an equilibrium point and eventually converge to the equilibrium point.

**global maximum** The largest element of a set or the largest value of a function over its entire range of observation, e.g., the largest value of the objective function used in solving an optimisation problem. **S**: absolute maximum

**global minimum** The smallest element of a set or the smallest value of a function over its entire range of observation, e.g., the smallest value of the objective function used in solving an optimisation problem. **S**: absolute minimum

**global navigation satellite system** A positioning system, which uses a constellation of satellites circumnavigating the Earth to provide the absolute position on the Earth surface, e.g., American Navstar GPS, European Galileo, Russian GLONASS, Chinese BDS. **S**: GNSS

**globe valve** A valve, which is designed to stop, start and control the flow of usually nonaggressive liquids, slurries, gases or vapours by moving the plug up and down. It is suitable for manual and automatic operation. It has a relatively large pressure drop in the fully open position and is commonly used in the piping of, e.g., cooling-water system, fuel-oil system, feed-water system, chemical-feed system.

**GNC** $\leftrightarrow$ guidance, navigation and control

**GNSS** $\leftrightarrow$ global navigation satellite system

**goodness-of-fit criterion** $\rightarrow$ model-fit criterion

**GPIB** $\leftrightarrow$ general-purpose interface bus

**gradient method** An algorithm that solves optimisation problems using search directions defined by the gradient of the function at the current point, e.g., gradient-descent method, conjugate-gradient method.

**graphical user interface** An environment, which simplifies the interaction between a user and a computational device by representing different options as graphical elements or visual indicators, e.g., icons, menus, buttons, dialogue boxes, scrollbars. By selecting one of these elements, using, e.g. mouse, keyboard, touch screen, the user can initiate various activities. It is used in industrial applications, as well as in smartphones, gaming devices, MP3 players, household, office. **S**: GUI

**graphic model** A symbolic model, which is given in the form of a drawing or a diagram

**grasp planning** The activity of determining where and how to grasp an object in order to provide a stable grasp.

**gravimetric humidity sensor 1**. A humidity sensor, which determines the absolute humidity from the difference between the weight of a piece of dry hygroscopic material and of the same piece of hygroscopic material exposed to the measured gas. **2**. A humidity sensor, which determines the absolute humidity from either surface acoustic-wave property changes caused by density changes induced by vapour absorption, or using piezoresistive micro-cantilever MEMS technologies.

**gravity compensation control** Control of a robot, which uses an additional control loop to calculate the gravitational forces from the actual robot position and directly adds them to the controller output. It is an upgrade of position control.

**grey-box model** A mathematical model that combines a theoretical structure with observation data to complete the model. The ratio between the theoretical-structure part and the part of the model obtained from data varies. **S**: hybrid model (3)

**grey noise** Coloured noise, the power spectral density of which is such that a human listener perceives that it is equally loud at all frequencies. Therefore, its power spectral density follows a psycho-acoustic curve of equal loudness.

**gripper** A mechanism, usually with two fingers, which can grasp objects of different shape, mass and material, e.g., robot gripper. It is driven by either a pneumatic actuator, a hydraulic actuator or an electrical actuator and can be equipped with force sensors and proximity sensors.

**ground mobile system** A mobile system equipped with wheels, caterpillars, legs mimicking human or animal walking, or with a mechanism mimicking some other type of animal locomotion, e.g., slithering. It can either carry an operator or act without one.

**GUI** ↔ graphical user interface

**guidance** The process of determining the desired path and changes in velocity, rotation and acceleration of a vehicle moving from the current location to a designated location.

**guidance, navigation and control** A subfield of control engineering which provides control-system design methods for the movement of vehicles. **S**: GNC

**guidance system** A device or a group of devices that calculates changes in position, velocity, altitude or angular velocity of the controlled mobile system when following the prescribed trajectory. It is often used in, e.g., ship, aircraft, spacecraft.

**guided robot** A robot, the movement of which is controlled using various technologies, e.g., hand-guided robot, GNSS-guided robot, CAD-guided robot, network-guided robot, vision-guided robot.

**gynoid** A robot, which is built to resemble a female human. **S**: fembot

**gyroscope** A sensor, which measures angle or angular velocity, e.g., ring-laser gyroscope, fibre-optic gyroscope, vibration gyroscope, fluid gyroscope. It is used for orientation and balance control in, e.g., aircraft, spacecraft, race car, motorboat, as well as for motion sensing. **S**: gyro sensor

**gyro sensor** → gyroscope

# Chapter 9
# H

**Hall-effect sensor** A distance sensor or proximity sensor, which is activated by an external magnetic field, generating analogue output as Hall voltage that is proportional to magnetic field density around the device. It can also act as a switch generating binary output. It is used in, e.g., proximity detection, distance detection, speed detection, flow measurement, temperature measurement, pressure measurement.

**Hammerstein model** A block-structured nonlinear mathematical model, which is made up of two submodels connected in series, where the first submodel is a static nonlinear mapping, which is connected to the input of a dynamic linear submodel. It is often combined with the Wiener model into various block structures.

**Hammerstein-Wiener model** A block-structured nonlinear mathematical model, which is made up of three submodels connected in series. The first submodel is a static nonlinear mapping, which is connected to the input of a dynamic linear submodel, the output of which is connected to the third submodel, again a static nonlinear mapping.

**hand coordinate frame** A Cartesian coordinate frame attached to the robot end-effector.

**hand-operated valve** $\rightarrow$ manual valve

**hand valve** $\rightarrow$ manual valve

**haptic display** A mechanical device, which transfers kinesthetic or tactile stimuli to the user. It provides the user with the feeling of touch, limited motion, compliance, friction, and texture in the virtual environment.

**haptic feedback** A mode of communication with the user, which enhances audio-visual communication, simulating the sense of touch. When the user touches the corresponding device, the latter can touch him back. It is usable in, e.g., education, training, industry, medicine, smart home, video games.

© ZRC SAZU/Research Centre of the Slovenian Academy of Sciences and Arts 2023    91
R. Karba et al., *Terminological Dictionary of Automatic Control, Systems and Robotics*,
Intelligent Systems, Control and Automation: Science and Engineering 104,
https://doi.org/10.1007/978-3-031-35755-8_9

**haptic interface** An interface, which enables the communication with computer or electronic device via sensory feedback. It comprises a haptic display, control software and power electronics, enabling the exchange of energy and forces between the user and the virtual environment.

**haptic rendering** A process of computing and displaying contact forces and tactile representations of real, remote or virtual objects in a computer-simulated environment. Force information from a computer algorithm is transmitted to an interface, which is capable of conveying information by enabling the user to touch, sense and manipulate objects in a virtual environment via a haptic interface.

**haptics** A technology, which transmits and understands information through touch. It uses tactile sensations in interfaces enabling touch and manipulation of objects.

**haptic simulation** Experimentation with a model of the environment, which enables telepresence using a haptic interface.

**hard valve seat** A valve seat made of metal or alloy, which usually cannot entirely prevent leakage in, e.g., check valve, globe valve, gate valve.

**hardware-in-the-loop simulation** A real-time simulation, which usually includes a plant simulator that simulates the controlled system as well as A/D converters and D/A converters enabling the connection with the tested hardware, e.g., PLC, microcomputer controller, embedded system. Hence, the hardware can be efficiently checked before its installation in the real control system.

**harmonic drive** A mechanism with a high transmission ratio, which uses inner and outer gear bands to provide smooth movement of the robot joint. It is compact and of low weight with no backlash.

**harmonic linearisation method** $\rightarrow$ describing-function method

**harmonic response** $\rightarrow$ frequency response

**HART protocol** $\leftrightarrow$ highway-addressable-remote-transducer protocol

**heading** $\rightarrow$ yaw angle

**heading, elevation, bank** $\rightarrow$ Tait-Bryan angles

**heat capacity** The parameter of a mathematical model, describing the property of an object to be able to store heat, defined as the ratio of the stored heat change to temperature change. It is a measure of the ability of the object to store heat. **S**: thermal capacitance, thermal capacity

**heat flow** Transfer of thermal energy trough an object, e.g., a heat sink, defined as the time derivative of the amount of energy being transferred.

**heat-flow density** $\rightarrow$ heat flux

**heat flow-rate intensity** → heat flux

**heat flux** Heat flow through material divided by the area, through which heat is conducted. **S**: heat-flow density, heat flow-rate intensity, thermal flux

**Heaviside function** → unit step function

**hierarchical control** Control of a large-scale system with a multilevel pyramid structure of the decision elements. The information flow is vertical, going from the fast data-sampling lower level towards the slower data-sampling upper level, and vice versa. However, the general control goal does not need to be the same as the control goals of the individual subsystems.

**high-cut filter** → low-pass filter

**high-performance liquid chromatograph** A chromatograph, in which a highly pressurised liquid solvent containing sample mixture is pumped through a column filled with solid adsorbent material, e.g., normal phase HPLC, reverse phase HPLC, size-exclusion HPLC, ion-exchange HPLC. Each component in the mixture interacts slightly differently with the adsorbent resulting in different flow rates of components at the end of the column. They are detected using various techniques, e.g., ultraviolet spectroscopy, mass spectrometry, fluorescence. It is used in, e.g., chemical separation and purification, pharmaceutical development and QC, drug and nutrient analysis, detection of illicit drugs in blood or urine, clinical testing, pollutant monitoring. **S**: HPLC

**highway-addressable-remote-transducer protocol** A master-slave fieldbus-based open protocol, which communicates bidirectionally over widely spread 4-20 mA analogue signal loops. It can have either a point-to-point or a multidrop topology. The former enables one device per loop, using analogue and digital connections, whereas the latter enables the communication of up to 63 devices, using an exclusively digital connection. It also provides wireless communication for process automation. **S**: HART protocol

**H-infinity** A design method for robust control, which provides a linear controller that assures closed-loop stability and guaranteed performance. The method is based on optimisation in Hardy space.

**HMI** ↔ human-machine interface (1, 2)

**HMI terminal** → operator panel

**hold equivalent** A method for the system discretisation, which results in a discrete-time model, the output of which in the sampling instants is the same as the output of the original continuous-time model, considering special input-signal related assumptions, e.g., step invariance, FOH equivalent.

**holding action** The operation of a controller, which keeps the control signal value constant until the next calculated value of the control signal becomes available. The same value can be constant through several sampling times, especially when a computationally complex control algorithm is implemented.

**holistic design** A design of a system as a whole, considering the interdependence of its parts or subsystems.

**holonomic system** A system, which has no kinematic constraints. Therefore all directions of motion in space are possible. Returning the internal variables of the system, such as angular displacement of the wheel or the robot joint angle, to the initial pose, the mobile robot or robot manipulator returns to its initial pose using the optional path.

**homogeneous transformation** A 4x4 transformation matrix, which encompasses 3x3 rotation matrix and 3x1 translation column. It describes the pose of a coordinate frame with respect to another coordinate frame or the translation and rotation of a coordinate frame.

**hot-wire anemometer** An anemometer, which measures heat loss of a wire placed in a flowstream using the constant-current method or the constant-temperature method. It consists of a sensing element, a thin electrically-heated wire exposed to the measured fluid flow, and electronic equipment generating the output signal. Proportionality with the measured flow rate is obtained from the resistance change caused by the temperature change of the wire, or from the current needed to maintain the temperature and thus the resistance of the wire unchanged. **S**: thermal anemometer

**household robot** A robot, which intelligently and autonomously moves itself or its parts within a domestic or similar environment to perform the intended task.

**HPLC** ↔ high-performance liquid chromatograph

**human-in-the-loop simulation** A real-time simulation, which involves human influence in simulated system behaviour using corresponding interfaces. It is useful in, e.g., training of operators, flight simulators, computer games. **S**: man-in-the-loop simulation

**human-machine interface** **1**. A structured combination of hardware and software, which is used to translate data from a control system into a human-readable visual representation of the controlled-system status, and to translate human interventions into the corresponding commands to a control system. It utilises one or several devices, e.g., keyboard, joystick, mouse, touch screen, operator panel, haptic interface as well as voice recognition, handwriting recognition, gesture recognition, brain-sensing devices. **S**: HMI, man-machine interface **2**. A communication platform, which allows the operator to control, monitor, collect data or program an industrial robot using different devices, e.g., teach pendant, computer. **S**: HMI, man-machine interface

# 9 H

**humanoid robot** A robot, which has physical properties of a human appearance, bipedal walking, manipulation and machine vision. It acts mostly in human environments interacting and communicating with humans. It has a high number of DOF, usually more than 30. **S**: android (1), anthropomorphic robot

**humidity sensor** A sensor, which measures the absolute or the relative amount of water in gases or solids, e.g., gravimetric humidity sensor, hair tension humidity sensor, chilled-mirror dew-point hygrometer, psychrometer, capacitive humidity sensor, resistive humidity sensor, thermal humidity sensor, optical humidity sensor. It is used in, e.g., industry, electronics, agriculture, medicine. **S**: hygrometer

**hybrid agent** An agent, which is a mixture of a deliberative agent and a reactive agent. It follows its plans but sometimes directly reacts to external events without deliberation to meet hard deadlines. It integrates different styles of reactive, deliberative and collaborative problem-solving in a modular fashion.

**hybrid computer** A computer, which combines the features and functionalities of an analogue computer as well as of a digital computer. It is usable in complex simulations, where calculation speed plays a key role.

**hybrid control** 1. Control, where the controller or the controlled system exhibits both continuous and discrete behaviour. 2. Control of robot end-effector position with simultaneous control of the contact force between the robot and the environment, i.e., the combination of position control and force control.

**hybrid model** 1. A mathematical model, which combines a continuous model with a discrete-event model. Its advantage is the possibility of encompassing a large class of systems uniformly in a flexible way, describing, e.g., continuous processes with discrete sensors, continuous processes with discrete actuators, batch processes. 2. A mathematical model, which is obtained by implementing different modelling methods. It is intended to improve the modelling accuracy or to enable modelling on specific problem domains, e.g., upgrade of the first-principles model with a fuzzy model or an ANN. **S**: integrated model 3. → grey-box model

**hybrid modelling** 1. Modelling of a dynamic system, which combines continuous and discrete-event behaviour. 2. A modelling approach that combines both first-principles modelling and system identification.

**hybrid simulation** A simulation, which combines a simulation environment and an experiment on a real system or an experiment using a virtual reality tool. It enables a realistic view of the problem under investigation.

**hybrid system** A dynamic system, which combines continuous and discrete-event behaviour. It is described by the corresponding hybrid model.

**hydraulic capacitance** 1. The parameter of a mathematical model, describing the property of a hydraulic assembly, e.g., storage tank, to store incompressible fluid, defined as the ratio of volumetric flow rate to the time derivative of fluid level. It is a measure of the ability to store liquid potential energy. **S**: hydraulic capacity **2**. The parameter of a mathematical model, describing the property of a hydraulic assembly, e.g., storage tank, to store incompressible fluid, defined as the ratio of mass flow rate to the time derivative of fluid level. It is a measure of the ability to store liquid potential energy. **S**: hydraulic capacity **3**. The parameter of a mathematical model, describing the property of a hydraulic assembly, e.g., storage tank, to store incompressible fluid, defined as the ratio of volumetric flow rate to the time derivative of pressure. It is a measure of the ability to store liquid potential energy. **S**: hydraulic capacity **4**. The parameter of a mathematical model, describing the property of a hydraulic assembly, e.g., storage tank, to store incompressible fluid, defined as the ratio of mass flow rate to the time derivative of pressure. It is a measure of the ability to store liquid potential energy. **S**: hydraulic capacity

**hydraulic capacity** → hydraulic capacitance (1, 2, 3, 4)

**hydraulic cylinder** A mechanical device, which converts power from the pressurised hydraulic fluid into linear motion, e.g., tie-rod cylinder, roundline cylinder, welded cylinder, telescopic cylinder. It comprises a moving piston, which is either single acting or double acting and generates strong forces and relatively slow movements. It is commonly used in, e.g., construction equipment, agricultural machinery, manufacturing machinery, mining, civil engineering. **S**: hydraulic piston

**hydraulic inertance** **1**. The parameter of a mathematical model, describing the property of moving incompressible fluid in a hydraulic assembly, e.g., pipe, defined as the ratio of pressure that accelerates or decelerates the fluid to the time derivative of volumetric flow rate. It is the measure of inertia of the liquid, which is moving in the pipe. **2**. The parameter of a mathematical model, describing the property of moving incompressible fluid in a hydraulic assembly, e.g., pipe, defined as the ratio of pressure that accelerates or decelerates the fluid to the time derivative of mass flow rate. It is the measure of inertia of the liquid, which is moving in the pipe.

**hydraulic motor** A mechanical device, which converts hydraulic pressure and hydraulic flow into torque and angular displacement, e.g., gear motor, vane motor, radial piston motor, axial piston motor. It operates with high starting-torque efficiency and is suitable for extreme environmental conditions. It is commonly used as a part of, e.g., conveyor, rolling mill, mobile robot, bulldozer, excavator.

**hydraulic piston** → hydraulic cylinder

**hydraulic resistance** **1**. The parameter of a mathematical model, describing the property of a hydraulic assembly, e.g., orifice plate, valve, to resist incompressible-fluid laminar flow. It is defined as the ratio of pressure drop to volumetric flow rate. **2**. The parameter of a mathematical model, describing the property of a hydraulic assembly, e.g., orifice plate, valve, to resist incompressible-fluid laminar flow. It is defined as the ratio of pressure drop to mass flow rate.

**hydrometer** A density meter, which utilises the equilibrium between the gravitational and buoyant forces, e.g., lactometer, alcoholometer, saccharometer, acidometer, urinometer. It consists of a sealed hollow glass tube attached to a glass bulb weighted with mercury or lead shot, which makes it float upright. The tested liquid is in a tall container where the glass tube floats freely. Where the surface of the liquid touches the glass tube, the measurement can be read from a calibrated scale. **S**: areometer

**hydrostatic level sensor** A level sensor, which uses a specifically configured pressure sensor to measure the height of fluid above the installation point of the sensor. The measured pressure is proportional to the height of the liquid with known density and local gravitational acceleration.

**hygrometer** $\rightarrow$ humidity sensor

**hyperparameter 1**. A parameter, which describes the distribution of parameter values of the analysed system. **S**: meta-parameter **2**. A parameter, which controls the learning process in machine learning. It can not be learned directly from the analysed data and must be therefore predefined. **S**: meta-parameter

**hyperredundant robot** A robot mechanism with a much higher DOF than needed concerning the task to be performed. It has usually the form of a snake, tentacle or elephant trunk.

**hysteresis** The difference between the response of a system to an increasing signal and its response to a decreasing signal. It is described by a nonlinear input-output characteristic. It appears in, e.g., thermostat, gear transmission, magnetisation.

# Chapter 10
# I

**IAE criterion** ↔ integral of the absolute value of the error criterion

**ICAD** ↔ individual channel analysis and design

**ICD** → individual channel analysis and design

**I-controller** ↔ integral controller

**identifiability 1.** The property of a system that it can be modelled using system identification. **S:** identificability **2.** The theoretical possibility to estimate the true values of the parameters of a system model after obtaining an infinite number of observations from the system. In practice, it enables the identification of accurate or close-to-true values of parameters of a mathematical model of a system after obtaining a large enough number of observations from the system. **S:** identificability

**identificability** → identifiability (1, 2)

**idle-channel noise** The noise in a communication channel that occurs under defined conditions when the information-carrying signal is not present.

**IEEE-488** → general-purpose interface bus

**if-part** → antecedent

**if-then rule** A two-part statement, which consists of the antecedent and the consequence and defines a logical relation. **S:** if-then statement

**if-then statement** → if-then rule

**IIoT** ↔ industrial internet of things

**ill-conditioned system** A system with a high condition number. Consequently, small changes in the input lead to a large change in the output. Such system is hardly controllable and is very sensitive to model uncertainties.

© ZRC SAZU/Research Centre of the Slovenian Academy of Sciences and Arts 2023
R. Karba et al., *Terminological Dictionary of Automatic Control, Systems and Robotics*,
Intelligent Systems, Control and Automation: Science and Engineering 104,
https://doi.org/10.1007/978-3-031-35755-8_10

**Ilon wheel** → Mecanum wheel

**imitation learning** A process, in which robot gets knowledge from a human teacher who shows how to execute the desired task. The robot than replicates human execution. It is appropriate for tasks in an unstructured environment. **S:** learning by demonstration, learning by watching, programming by demonstration

**imitation model** A physical model with static character, which emphasises specific aspects of the modelled system, e.g., doll, template, atomic structure.

**impedance control** Control of the deviation of the robot end-effector motion from the desired motion due to the interaction with the environment, which is related to the contact force. The method is used in collaborative operations and haptic interfaces.

**implantable technology** A set of devices, which can be permanently or semi-permanently embedded in a human body providing benefits that cannot be afforded by non-embedded devices, e.g., health monitoring, wellness monitoring, precise drug delivery, disease monitoring. Various tiny electronic devices, e.g., identity implant, biodegradable bio-battery, implantable birth control, healing chip, cyber pill, brain-computer interface, are constrained by a set of specific requirements, e.g., biocompatibility, bio-resistance, size, density matched to neural tissue, minimal tethering to adjacent structures. **S:** embeddable technology

**implicit function generation** → implicit method

**implicit method** A method, which enables the derivation of a simulation scheme from a function with time as the independent variable. The successive derivations of the function lead to a differential equation, the solution of which is the simulated function. The derived differential equation is then solved with the indirect method. **S:** implicit function generation

**improper transfer function** A transfer function, in which the degree of the numerator exceeds the degree of the denominator. Its relative degree is less than 0. It describes a system that is neither causal nor stable.

**impulse response** The response of a system to an input signal, described with the Dirac delta function. For a linear time-invariant system, it is defined for all frequencies.

**IMU** ↔ inertial measurement unit

**INA method** ↔ inverse-Nyquist-array method

**inclinometer** A sensor, which measures the positive or negative slope of an object with respect to the direction of gravity, given in degrees or percentages. It is used in, e.g., robotics, aviation, nautical systems, security systems as well as in, e.g., game controller, smartphone, digital camera. **S:** roll-pitch sensor, tilt sensor

**incremental encoder** **1.** An encoder, which measures relative changes in position from the point of its initialisation, as well as their direction, e.g., optical encoder, magnetic encoder. It generates a pulse for each step, which provides the corresponding electrical output. It needs to be powered on throughout the entire measurement cycle. It is commonly used in applications that require precise measurement and control of position or velocity also in demanding and harsh environments. **S:** relative encoder **2.** An encoder, which measures relative changes in orientation from the angle of its initialisation, as well as their direction, e.g., optical encoder, magnetic encoder. It generates a pulse for each step, which provides the corresponding electrical output. It needs to be powered on throughout the entire measurement cycle. It is commonly used in applications that require precise measurement and control of orientation or angular velocity also in demanding and harsh environments. **S:** relative encoder

**indirect method** **1.** A method, which enables the derivation of a simulation scheme from a mathematical model described by a differential equation. The highest order derivative of the latter must be explicitly expressible and the differential equation must not include derivatives of the input. The differential equation is solved by integrating the highest order derivative as many times as its order. **2.** A method, which enables the derivation of a simulation scheme from a mathematical model described by a difference equation. The predictive output variable must be explicitly expressible and the difference equation must not include delayed input samples. The difference equation is solved by delaying the sample of output variable as many times as its order.

**individual channel analysis and design** A methodological framework for the analysis and synthesis of MIMO systems. The latter are treated as the sets of SISO subsystems, while the cross-couplings are considered as disturbances. It also enables the robustness analysis and robust control design. **S:** ICAD, ICD, individual channel design

**individual channel design** → individual channel analysis and design

**induction motor** An AC motor, the rotation of which is not in phase with the rotating magnetic field of the stator. The latter generates a current in the squirrel cage or the wound-type rotor, in turn generating a force, which draws the rotor towards the moving magnetic field of the stator. It is commonly used as a part of, e.g., fan, blower, escalator, electrical vehicle, process control system, but especially in the case of heavy or high-speed workload. **S:** asynchronous motor

**inductive displacement sensor** **1.** A displacement sensor, which utilises changes in output voltage when a movable magnetic core, connected with the measured movement, is shifted into a coil, e.g., LVDT. **2.** A noncontact, compact, low-cost displacement sensor, which measures small distances between the sensing element and the target object or material. When the distance changes, eddy currents are induced in a conductive object resulting in an appropriate output signal. It is commonly used in industry due to its high resolution and tolerance to dirty environments. **S:** eddy current displacement sensor

**inductive level sensor** A level sensor, which consists of a tube containing the measured fluid, where a change of its level affects the impedance of the coil that surrounds it or a conductive fluid changes the resistance of the coil. It is usable in harsh and dirty environmental conditions.

**inductive proximity sensor** A compact, low-cost proximity sensor, which induces eddy currents in a conductive object. The detection of eddy currents triggers a change of its binary output. **S:** eddy current proximity sensor

**inductive reasoning** The reasoning about broad generalisations on the basis of specific observations.

**industrial blower** A mechanical device, which is used to provide and accommodate a large flow of air or gas to a specific area in the room or space. It consists of a wheel with small blades, which draws medium into the inlet of housing that directs flow out. It has heavier construction to meet the demands of industrial applications, creating suction, pressurisation, cooling, combustion air, forced ventilation or general ventilation in, e.g., dust collection, silo loading, blow-off systems, food processing.

**industrial controller** A controller specialised for manufacturing and process control in extreme environments, e.g., PLC, PAC. It is a high performance, complete ready-to-employ system, which include also bumpless transfer, anti-windup and signal converter. It offers a variety of structures, such as on-off control, PID control, fuzzy control, self-tuning, manual tuning. **S:** process controller

**industrial Ethernet** An Ethernet-based communication protocol, which operates in a harsh environment, enabling determinism and real-time control, e.g., EtherCAT, EtherNet/IP, PROFINET, CC-Link IE, Modbus. It provides predictable performance, maintainability and plug-and-play interoperability among complex devices from multiple vendors.

**industrial fan** A mechanical device, which is used to provide and accommodate a large flow of air or gas around the room or space. It can be either centrifugal or axial. It consists of vanes or blades that rotate in a housing, increasing pressure and adding kinetic energy to the medium. It has heavier construction to meet the demands of industrial applications, e.g., chemical processing, corrosive gas handling, dust collection, fume control, process cooling and heating.

**industrial internet of things** An integral part of a modern manufacturing concept and an extension of the internet of things. It enables connectivity, which increases the overall efficiency, and reduces human errors and manual labour, resulting in cost savings and time savings. It is used in, e.g., robotics, manufacturing, automotive industry, oil and gas industry, smart grid, smart city, and is tightly connected with various technologies, e.g., cloud computing, cognitive computing, machine-to-machine communication, mobile technologies, 3D printing. **S:** IIoT

**industrial PC** A PC intended and adapted to the harsh operational environment. It has improved reliability, maintainability, durability, precision standards, safety and security to cope with the demands of the industrial environment. It is usually used for process control and data acquisition.

**industrial robot** A feedback-controlled, programmable, multipurpose robot, which is used in manufacturing. It can have three or more DOF.

**inertial measurement unit** A sensor-fusion device, which reports data related to orientation, velocity and gravity, thus combining a gyroscope, accelerometer and magnetometer. It is used in, e.g., mobile robot, rocket, smartphone, pedometer. **S:** IMU

**inertial navigation** A dead reckoning, which bases on direct measurements of inertia.

**inertial navigation system** A system, which uses inertial sensors to determine the position, orientation, or velocity of a mobile system or of another moving device.

**inertial sensor** A sensor for measuring the changes in orientation or velocity of a mobile system or of another moving device, e.g., accelerometer, gyroscope. It utilises linear momentum balance and angular momentum balance or the measurement of the forces and torques generated by the changes in linear momentum and angular momentum.

**infrared camera** → thermal imaging camera

**infrared pyrometer** A pyrometer, which measures the thermal radiation emitted by the measured object, e.g., wideband infrared pyrometer, narrowband infrared pyrometer, two-colour infrared pyrometer. It consists of a lens, which focuses the incoming infrared thermal radiation onto the black body, the temperature of which is measured by a corresponding sensor, e.g., thermocouple, thermistor, RTD, and selective filter. It enables contactless measurements of a wide range of temperatures in various applications.

**infrared sensor** A noncontact sensor, which is used in wireless remote control and detection of surrounding objects. It consists of a LED or a laser acting as a transmitter, and a photodiode or a phototransistor as a receiver. The transmission medium between the sensor and the object is usually vacuum, atmosphere, or optical fibres. Further processing of the received radiation energy indicates the required information. An active device contains both the transmitter and the receiver, while a passive device is only a receiver, whereas the measured object is the source of radiation. It is often used for monitoring and control applications.

**infrared thermometer** A noncontact temperature sensor, which infers the measured temperature from a portion of thermal radiation of the object. It consists of a lens, which focuses the infrared thermal radiation onto a detector. The detector is often a thermopile that converts the radiant power to an electrical signal proportional to the measured temperature. It is portable and enables immediate and accurate temperature readings for the stationary or moving objects in, e.g., engineering applications, medical applications.

**inherent regulation** $\rightarrow$ self-regulation

**inherent valve characteristic** A graph of the percent of the flow rate against the valve travel, which is expressed as the percent of the valve throat area, e.g., linear valve characteristic, equal percentage valve characteristic, quick-opening valve characteristic, modified parabolic valve characteristic, square-root valve characteristic. It is obtained on a standardised test line at a fixed pressure drop, as defined for the determination of the flow factor or flow coefficient. It is typically provided by valve manufacturers.

**input decoupling zero** The decoupling zero of a linear MIMO system representing an uncontrollable system mode.

**input disturbance** A disturbance, which is added to the control variable at the point of the controlled system where the actuation takes place. It is mainly the consequence of external causes resulting in perturbations of the control signal.

**input error** The difference between the system input signal and the output signal of the inverse model, where the system output signal is the input signal of the inverse model.

**input-error model** $\rightarrow$ inverse model

**input matrix** A matrix in the vector-matrix form of linear state equations of MIMO system, usually marked with letter $B$, which defines the relations between the system inputs and the state variables. **S:** control matrix

**input-output error** $\rightarrow$ equation error

**input-output model** A mathematical model, which describes the relations between the selected input and output of the considered system, e.g., differential equation, transfer function.

**input-output sensitivity** The ratio between the change of the output variable and the change of the input variable, which causes the output change.

**input-output stability** $\rightarrow$ bounded-input bounded-output stability

**input sensitivity function** $\rightarrow$ load disturbance sensitivity function

**input signal** A signal that enters a system, e.g., the excitation of a dynamic system.

**input variable** A variable that excites or disturbs the system-output response.

**input vector 1.** A vector of all the variables that enter the mathematical model of a MIMO system. **S:** control vector **2.** A vector in the vector-matrix form of linear state equations of a SISO system or of a SIMO system, which defines the relations between the system input and the state variables. It is usually marked with the letter $b$. **3.** A vector, which consists of all system inputs. In the case of only one input, it degenerates into a scalar.

**installed valve characteristic** A graph of the flow rate against the valve travel, measured at the actual pressure drops. In most applications, the pressure drop increases as the flow rate decreases. Therefore, it normally changes from an equal-percentage valve characteristic to a linear valve characteristic and from a linear valve characteristic to a quick-opening valve characteristic.

**instruction list** A textual standard PLC programming language that implements a program as an ordered set of low-level commands. **S:** statement list

**integral-action time** $\rightarrow$ integral time

**integral control** A control strategy, in which the rate of change of the controller output is proportional to the error. Hence, the controller output is proportional to the integral of the error with respect to time. It eliminates the steady-state error. On the other hand, it responds slowly. **S:** reset control

**integral controller** A controller, the output of which is the integral of the error multiplied by a constant. Its input is the error, whereas its output is the control signal. It eliminates steady-state error, but it is rarely used alone. **S:** I controller, reset controller

**integral criterion** An objective function used for quantitatively measuring the performance of a control system. It includes some form of the error signal, usually the integral of the absolute value of the error or the integral of the square of the error, e.g., IAE criterion, ISE criterion, ITAE criterion, ITSE criterion, ISTAE criterion, ISTSE criterion.

**integral gain** A constant, which is the factor between controller output and integral of the error. It is used in feedback control with the controller that contains an integral term. It is given also as the quotient of proportional gain and integral time. **S:** integration gain

**integral of squared time multiplied by the absolute value of the error criterion** An objective function, which is calculated as the integral of the product of the square of the time and the absolute value of the error. It is used for quantitatively measuring the performance of a control system by considering the absolute amplitude of the error signal with further emphasis on long-duration errors or steady-state errors due to the square-of-the-time multiplier of the error function. Therefore, it is often applied in systems requiring an extremely fast settling time. **S:** ISTAE criterion

**integral of squared time multiplied by the squared error criterion** An objective function, which is calculated as the integral of the product of the square of the time and the square of the error. It is used for quantitatively measuring the performance of a control system by considering the energy of the error signal with further emphasis on long-duration errors or steady-state errors due to the square-of-the-time multiplier of the error function. Therefore, it is often applied in systems requiring an extremely fast settling time. **S:** ISTSE criterion

**integral of the absolute value of the error criterion** An objective function, which is calculated as the integral of the absolute value of the error. It is used for quantitatively measuring the performance of a control system by considering the absolute amplitude of the error signal. **S:** IAE criterion

**integral of the squared error criterion** An objective function, which is calculated as the integral of the square of the error. It is used for quantitatively measuring the performance of a control system by considering the energy of the error signal. As the square value of an error function is often analytical, it facilitates the analysis. **S:** ISE criterion

**integral of time multiplied by the absolute value of the error criterion** An objective function, which is calculated as the integral of the product of the time and the absolute value of the error. It is used for quantitatively measuring the performance of a control system by considering the absolute amplitude of the error signal with an emphasis on long-duration errors or steady-state errors due to the time multiplier of the error function. Therefore, it is often applied in systems requiring a fast settling time. **S:** ITAE criterion

**integral of time multiplied by the squared error criterion** An objective function, which is calculated as the integral of the product of the time and the square of the error. It is used for quantitatively measuring the performance of a control system by considering the energy of the error signal with the emphasis on long-duration errors or steady-state errors due to the time multiplier of the error function. Therefore, it is often applied in systems requiring a fast settling time. **S:** ITSE criterion

**integral term** A subsystem, the output of which is is obtained by integrating its input signal and multiplying the result by a constant. It is usually a constituent part of a controller. **S:** I term

**integral time** A coefficient, which determines the impact of the I term of a controller on closed-loop performance. It is equal to the time, in which the share of the I term is equalised with the share of the P term in the step response of a PI controller. **S:** integral-action time, reset time

**integral windup** A phenomenon, which occurs in a control system that includes a controller with an I term. An unreachable setpoint causes saturation of the final control element or the actuator. Therefore, the error keeps accumulating and the I term increases. This results in a nonresponsive controller or a high overshoot of the controlled signal, which can occur even after the saturation ceases. **S:** reset windup.

**integrated-circuit temperature sensor** A temperature sensor, which utilises the temperature dependence of a p-n junction of a diode or a base-emitter junction of a transistor, the output current of which is proportional to the absolute temperature. It is accurate, linear, reliable and often a part of a circuit board to monitor and control the temperature in, e.g., communication applications, special industrial applications, overheating protection applications. **S:** silicon bandgap temperature sensor, solid-state temperature sensor

**integrated manufacturing system** A group of machines, which work together in a coordinated manner. They are linked by a material-handling system and correspondingly controlled. It is used for production, treatment, movement or packaging of discrete parts or assemblies.

**integrated model** → hybrid model (2)

**integrated PLC** → compact PLC

**integrated system** A complex multidomain system, which needs concurrent, multidisciplinary, model-dependent design procedure, using virtual prototyping.

**integrating system** A system, the dynamics of which has at least one pole equal to 0, or at least one eigenvalue of the system matrix equal to 0, or at least one root of the characteristic equation equal to 0. Its step response is monotonically increasing or decreasing with the absolute value approaching infinity.

**integration gain** → integral gain

**integrator 1.** The element of a system, the output of which is the integral of the input signal. **2.** The block of a simulation scheme, the output of which is the integral of the input signal. The initial condition is taken into account at the start of simulation run.

**integrity** The property of a MIMO system, such that its stability is maintained in the case of the breakdown or failure of a component in a single loop.

**intelligent building** → smart building

**intelligent control** A set of control techniques, which comprises various methods from the field of artificial intelligence, e.g., fuzzy control, expert systems, reconfigurable control. It is used in, e.g., nonlinear-system control, model-free control, data-based nonlinear estimation.

**intelligent sensor** A smart sensor, upgraded with additional functions, which can learn, judge or process signals, can self-test, self-validate, self-adapt and can detect conditions and respond to them. Therefore, its accuracy, resolution, stability, reliability and adaptability are improved. Its built-in knowledge enables a certain level of reasoning and decision-making. It is used in, e.g., industrial application, medical application.

**intelligent vehicle** A vehicle, which is enhanced with perception, reasoning, and actuating devices that enable the automation of driving tasks such as safe line following, obstacle avoidance, overtaking slower traffic, following the vehicle ahead, assessing and avoiding dangerous situations, and determining the route.

**intentional agent** → deliberative agent

**interacting system** → interactive system

**interaction** → cross coupling

**interactive system** A MIMO system with cross-couplings, which are not negligible in the process of system analysis and control design. **S:** interacting system

**interface** A hardware or software equipment, which enables the connection of two or more incompatible systems or parts of a system and their mutual operation. It may also include a human operator.

**internal-model control** The control, which uses an explicitly or implicitly implemented model of the controlled system to determine the control signal.

**internal stability** The property of a system, the states of which remain bounded for all bounded initial conditions as well as for all bounded excitation signals that could influence the system at any place. This requirement concerns all states, regardless of their controllability or observability.

**interoperability** Ability of different information systems, devices and applications to access, exchange, integrate and cooperatively use data in a coordinated manner. Portability of information is ensured by the ability of equipment of different suppliers to function in a network, by the ability to exchange information among computers through local or wide area networks, and by the ability to share data among components or devices via software and hardware.

**interpolation function** → membership function

**interpreter-oriented simulation language** A simulation language, which includes the interpreter program that compiles and implements commands in a corresponding sequence. The user can change the structure and parameters of a mathematical model simultaneously without additional compilation. It is convenient in the phase of model development.

**intersection point** → centroid

**interval-oriented simulation** → discrete-time simulation

**invariant zero** The zero from a set of transmission zeroes and some decoupling zeros of a linear MIMO system that does not change when either the state feedback or the output feedback is applied. In the case of a minimal realisation of a description of a linear MIMO system. it is equal to the transmission zero.

**inventory-management system**  A combination of hardware and software as well as processes and procedures that oversee monitoring, ordering, storing, using and maintenance of company assets, raw materials, supplies or finished products. It enables, e.g., better organisation, improved cash flow, better reporting and forecasting capabilities, tracking and transparency, reduction in deadstock, improved supplier, vendor and partner relationship. It is used in, e.g., industry, healthcare, government.

**inverse dynamic model**  A mathematical model, which enables the calculation of the robot joint forces and torques that generate the desired robot end-effector trajectory or enables limb-motion analysis. It is used in, e.g., robot control, path planning, biomechanics.

**inverse dynamics control**  Control of the robot joint variables applying a feedback linearisation to a nonlinear system. The method causes cancelling of the nonlinear terms and decoupling of the dynamics for each robot link. **S:** computed torque control

**inverse $f$ noise**  $\rightarrow$ pink noise

**inverse Jacobian control**  Control of a robot, which relates the robot end-effector velocities with the robot joint velocities. The control principle describes the behaviour of a spring expressed in robot joint variables.

**inverse kinematics**  Kinematics, in which the robot-joint variables are calculated from the known robot end-effector pose. A similar calculation can also be carried out for robot-joint velocity and acceleration.

**inverse model**  A mathematical model where the inputs and outputs are swapped. Its identification is based on the prediction error of an input obtained from past values of the inputs and outputs. **S:** input-error model

**inverse-model-based control**  **1.** The control, which uses the inverse model of the controlled system to determine the control signal. The input of the inverse model is thus the reference signal, whereas the output is the feedforward control signal. **2.** The control, which uses the inverse model of the controlled system to compensate for some dynamic properties of the controlled system, usually nonlinearities.

**inverse-Nyquist-array method**  A MNA method that attenuates cross-couplings according to the inverse-Nyquist-diagram graphical criterion for the evaluation of diagonal dominance. The corresponding MIMO controller can be implemented with simple and commonly used control components. The obtained system is considered to be decoupled. Therefore, the design of additional SISO controllers for the individual subsystems is necessary. **S:** INA method

**inverse Nyquist diagram**  $\rightarrow$ inverse Nyquist plot

**inverse Nyquist plot**  A Nyquist plot for which the Nyquist contour is mapped over the inverse of the open-loop transfer function. **S:** inverse Nyquist diagram

**inverse problem I 1.** The problem of determining the dynamics of a system from the known input signal and output signal. Such a problem is often solved using identification. **2.** The problem of determining the state of a system from the known input signal and output signal. Such a problem is often solved using state estimation.

**inverse problem II** The problem of determining the input signal, which ensures as good as possible output-signal tracking of the prescribed reference trajectory. Such a problem is often solved by an appropriate control design.

**inverter** An electrical device, which converts DC power from, e.g., battery, solar cell, fuel cell, to AC power. It is used in, e.g., actuator system, UPS, induction heating, power grid.

**inverter drive** → variable-speed drive

**ionisation gauge** A high-vacuum sensor, which determines the pressure from the current between two electrodes of the triode-like device at a prespecified voltage. The conductivity of the gas is ensured by the electrons emitted from the cathode filament.

**i/p converter** → current-to-pressure converter

**irreversible process** A process, which proceeds in one direction, with the finite gradient between the states of the process. It is the direction of rising entropy, while the reverse direction is not possible or requires additional energy, e.g., diffusion.

**ISE criterion** ↔ integral of the squared error criterion

**isolated system** → closed system

**ISTAE criterion** ↔ integral of squared time multiplied by the absolute value of the error criterion

**ISTSE criterion** ↔ integral of squared time multiplied by the squared error criterion

**ITAE criterion** ↔ integral of time multiplied by the absolute value of the error criterion

**I-term** ↔ integral term

**ITSE criterion** ↔ integral of time multiplied by the squared error criterion

# Chapter 11
# J

**Jacobian matrix** A matrix, which contains partial derivatives of the robot end-effector coordinates with respect to the robot joint variables.

**jamming** The state, in which the insertion of the pin into a hole is unsuccessful due to unbalanced forces and torques acting on the robot.

**jet-pipe amplifier** A hydromechanical converter, which consists of a jet pipe supplied with a high-pressure hydraulic fluid that pivots about a fixed point. The jet pipe distributes the emitted fluid, accelerated through a nozzle, to two collector nozzles. The deflection of the pipe determines the position of the load, e.g., piston in a cylinder. It is typically used as a preamplifier of a multistage hydraulic amplifier.

**Jordan canonical form** A block-diagonal canonical form of the system matrix in the state-space model, which is used when the system matrix has multiple eigenvalues. **S:** Jordan normal form

**Jordan normal form** → Jordan canonical form

**Jury stability criterion** A method for analysing the stability of a linear time-invariant closed-loop SISO discrete-time system, where the stability depends on the number of poles outside the unit circle, which is determined using a table that is based on the coefficients of the characteristic equation.

© ZRC SAZU/Research Centre of the Slovenian Academy of Sciences and Arts 2023
R. Karba et al., *Terminological Dictionary of Automatic Control, Systems and Robotics*,
Intelligent Systems, Control and Automation: Science and Engineering 104,
https://doi.org/10.1007/978-3-031-35755-8_11

# Chapter 12
# K

**Kalman canonical decomposition** $\rightarrow$ Kalman decomposition

**Kalman decomposition** A structural decomposition that results in the corresponding canonical form of the mathematical-model description in state space. It enables a clear view of the controllable and observable parts of the system, which simplifies the analysis and control design of the system. **S**: Kalman canonical decomposition

**Kalman filter** An algorithm that estimates the unknown variables by estimating a joint probability distribution over the variables for each time sample. It uses a series of measurements over time, containing noise and other inaccuracies. It is a linear estimator in the minimum mean-square-error sense.

**key performance indicator** A quantifiable nonfinancial measure, which demonstrates how effectively a company is achieving operational objectives and evaluates the success at reaching targets. It is usable at different levels, focusing either on overall performance or it can be adapted to a considered department, e.g., sales, marketing, support. It can be adjusted for various purposes, e.g., manufacturing, system operation, professional service, project execution, supply-chain management, human-resource management. **S**: KPI

**K-factor** $\rightarrow$ thermal conductivity

**kill switch** $\rightarrow$ emergency stop

**kinematic chain** An assembly of rigid bodies, e.g., robot links, which are connected by joints, e.g., robot joints.

**kinematic model** A mathematical model, which describes the motion of a robot system regardless of forces and torques that affect the motion, e.g., the relation between trajectories, velocities and accelerations of robot joints and robot end-effector, giving its pose as a function of robot-joint variables.

© ZRC SAZU/Research Centre of the Slovenian Academy of Sciences and Arts 2023
R. Karba et al., *Terminological Dictionary of Automatic Control, Systems and Robotics*,
Intelligent Systems, Control and Automation: Science and Engineering 104,
https://doi.org/10.1007/978-3-031-35755-8_12

**kinematic pair** A basic element of a kinematic chain, which consists of two neighbouring robot links connected by a prismatic or rotational joint.

**kinematic singularity** A point in the robot workspace where the Jacobian matrix, which relates robot joint velocities to robot end-effector velocities, loses its rank. Consequently, it is not possible to invert the Jacobian matrix. A change of robot joint variables does not result in a change of robot end-effector pose. At this point, the mobility of the kinematic structure is reduced in serial robots and increased in parallel robots. For such a point, an infinite number of solutions to the inverse-kinematics problem may exist.

**kinematic structure** The physical composition of a robot system, including robot joints, robot links, and robot end-effector.

**kinesthetic sense** $\rightarrow$ proprioception (1, 2)

**KPI** $\leftrightarrow$ key performance indicator

**Kronecker delta function** A function defined on a discrete domain that is 0 everywhere but at the origin of the coordinate system, where it is 1. It is a discrete-time analogue of the Dirac delta function.

**K-value** $\rightarrow$ thermal conductivity

# Chapter 13
# L

**laboratory apparatus** $\rightarrow$ laboratory setup

**laboratory mockup** $\rightarrow$ laboratory setup

**laboratory pilot plant** $\rightarrow$ laboratory setup

**laboratory setup** A pilot plant, which is designed to be used in the learning process, taking place in a laboratory. It enables students to understand the addressed issue better and easier. **S**: laboratory apparatus, laboratory mockup, laboratory pilot plant

**ladder diagram** A graphical standard PLC programming language that depicts the connections between contacts and coils of logic relays.

**lag compensation** A design method for modifying the open-loop transfer function by introducing a phase delay, which changes the steady state of the closed-loop system. It provides a suitable phase margin by attenuating mid-frequency and high-frequency bands. It has little impact on the transient response of the closed-loop system. **S**: phase-lag compensation

**lag compensator** A compensator, the pole of which is closer to the origin of the $s$-plane than its zero. Therefore, it introduces a phase delay in the open-loop transfer function. **S**: phase-lag compensator

**lambda probe** $\rightarrow$ lambda sensor

**lambda sensor** A gas sensor, which measures the concentration of oxygen in the analysed gas or liquid, e.g., zirconia oxygen sensor, electrochemical oxygen sensor, optical oxygen sensor, infrared oxygen sensor. It is used in, e.g., fuel mixture feedback control, internal-combustion engine, fire-prevention system, diving equipment. **S**: lambda probe

**Laplace domain** $\rightarrow$ $s$-domain

© ZRC SAZU/Research Centre of the Slovenian Academy of Sciences and Arts 2023
R. Karba et al., *Terminological Dictionary of Automatic Control, Systems and Robotics*,
Intelligent Systems, Control and Automation: Science and Engineering 104,
https://doi.org/10.1007/978-3-031-35755-8_13

**large-scale system** A complex system, which can be decoupled or partitioned into less complex subsystems that affect each other. They are interconnected so that they compose a hierarchical system.

**laser distance sensor** A distance sensor, which uses the light beam time-of-flight principle to determine the distance to the measured object or its dimensions. It is used in, e.g., 3D object recognition, mobile robotics, industry, military, sports.

**laser interferometer** A precise displacement sensor, which consists of a laser beam, complex optics and a photodetector that records the interference-pattern changes enabling the determination of the distance change to the measured object. It is used in, e.g., positioning, semiconductor metrology, lithography process in the production of integrated circuits, and due to its high precision also for calibration of machine tools, laboratory instruments and other displacement sensors. **S**: Michelson interferometer

**laser thermometer** **1**. A noncontact temperature sensor, which measures the vibration velocity of molecules of the medium around the measured object. The heat from the measured object influences the velocity changes of the molecules. Using the emitted light beam, these changes can be measured and converted to temperature. For an accurate measurement, the path of the beam must not be obstructed and the influence of the surrounding objects heat must be minimised. **2**. An infrared thermometer, which uses a light beam to help accurately point the device towards the measured object.

**law of conservation of energy** $\rightarrow$ energy balance

**law of conservation of mass** $\rightarrow$ mass balance

**law of parsimony** $\rightarrow$ Occam's razor

**lead compensation** A design method for modifying the open-loop transfer function by introducing a phase advance, which improves the relative stability of the closed-loop system. It significantly influences the transient response and has little impact on the steady state of the closed-loop system. **S**: phase-lead compensation

**lead compensator** A compensator, the zero of which is closer to the origin of the $s$-plane than its pole. Therefore, it introduces a phase advance in the open-loop transfer function. **S**: phase-lead compensator

**lead-lag compensation** A design method for modifying the open-loop transfer function by introducing a phase advance in one frequency band, and a phase delay in another frequency band.

**lead-lag compensator** A two-part compensator, which has the pole-zero pair of the phase-delay part closer to the origin of the $s$-plane than the zero-pole pair of the phase-advance part. Therefore, it introduces a phase advance in one frequency band and a phase delay in another frequency band of the open-loop transfer function, thus influencing transient response and steady state of the closed-loop system.

**lean enterprise**  → lean manufacturing

**lean manufacturing**  A production that guarantees the delivery of the required material or products at the prescribed location, time and quantity with minimal waste and minimal storage. At the same time, the flexibility of production and the quality and value of products are maintained with the least invested work, reduced production time and reduced costs. **S**: lean enterprise, lean production

**lean production**  → lean manufacturing

**learning by demonstration**  → imitation learning

**learning by watching**  → imitation learning

**learning control system**  A control system, which contains sufficient computational ability to develop the mathematical model of the controlled process, using feedback data and information about its environmental impact in the past. Its operation can be correspondingly modified taking advantage of the newly acquired knowledge to improve system behaviour in the future. Therefore, it is used in the control of problematic systems, e.g., control of poorly-modelled nonlinear systems.

**least-squares method**  **1**. A method for the optimisation of the mathematical model parameters, which uses an algorithmic search for the smallest sum of the squares of the differences between the response of the model and the measured response of the system. **2**. An approach in regression analysis to approximate the solution of an overdetermined system of equations.

**level**  → stock

**level gauge**  → level sensor

**level sensor**  A sensor, which measures the position of the border between two media with different densities relative to a horizontal plane, most often the border between a liquid and a gas relative to the bottom of a tank, e.g., capacitive level sensor, conductivity level sensor, ultrasonic level sensor, optical level sensor. Continuous or point measurements in open or closed systems with different fluids and fluidised solids including slurries, granular materials and powders are enabled. It is used in, e.g., pharmaceutical industry, food industry, beverage industry, wastewater treatment. **S**: level gauge

**level switch**  A device, which senses the presence of a medium at the place of installation and changes its binary output at the moment when the prescribed level is reached, e.g., vibrating fork, float switch, rotating switch, ultrasonic switch. It is used for the detection of the presence of various materials from bulk solids to liquids and everything in between, as well as for point-level monitoring.

**limit cycle**  A closed trajectory in phase space with the property that at least one other trajectory of dynamic-system response spirals into it when the time approaches infinity. It is a trajectory that represents the solution of certain types of nonlinear differential equations.

**limiter** An electrical device, which cuts off the signal that exceeds the prescribed threshold, while the signal below the prescribed threshold passes unaffected.

**limit switch** An industrial control component, which is operated by the motion of a machine part or presence of an object by closing or opening the contact when pressed with sufficient force. It can be of lever roller type or roller plunger type. It has a simple structure and consumes little electrical energy. It is utilised in industrial control applications, e.g., for safety interlocks, for counting objects passing a point, for triggering events. **S**: end-switch

**linear acceleration** The time derivative of linear velocity of an object.

**linear approximation** A graphic linearisation method in which the known static characteristics of the discussed system is replaced with the tangent in the chosen operating point. Such linearisation is justified only in the previously specified vicinity of the operating point. **S**: small signal approximation, tangent-line approximation

**linear displacement** The change of the position of an object along the shortest path between the initial point and the final point of such linear motion.

**linear encoder** **1**. An absolute encoder with a head, which measures position along a straight line, e.g., capacitive encoder, optical encoder, magnetic encoder. It is commonly used in, e.g., robotics, automation systems, machining tools. **2**. An incremental encoder with a head, which measures position along a straight line, e.g., capacitive encoder, optical encoder, magnetic encoder. It is commonly used in, e.g., robotics, automation systems, machining tools.

**linear induction motor** An induction motor, which directly generates linear displacement. Its primary coil consists of three-phase windings assembled on a steel lamination stack, while its secondary coil is an aluminium plate or a copper plate. The fixed primary coil is excited by the three-phase power supply causing a travelling flux along with the primary coil. The latter pulls the moving secondary coil along with the primary coil producing the required displacement. It offers continuous thrust forces, high speed and acceleration. It is commonly used in, e.g., machine tool, mechanical handling, conveying.

**linearisation** A procedure, which enables a nonlinear system to be approximated by a linear mathematical model in the neighbourhood of the selected operating point, e.g., linear approximation, Taylor-series expansion method, describing-function method.

**linearity** **1**. The property of a mathematical relation, which is additive and homogenous. **2**. A mathematical relation, the graphical representation of which is a straight line. **3**. The extent, to which the measured static characteristic of a sensor fits its linearised characteristic. It is often given as the ratio between the maximum input deviation and the measuring range expressed as a percentage.

**linear matrix inequality** A mathematical inequality used for the description of constraints. It is often used in certain convex optimisation problems in control theory. **S**: LMI

**linear model** A mathematical model described with linear mathematical structures, e.g., with linear differential equations. The law of superposition can be applied.

**linear motion** The movement of an object along a straight line in one spatial dimension. **S**: rectilinear motion, translational motion

**linear quadratic Gaussian controller** A controller that is a combination of a Kalman filter, which is a linear–quadratic state estimator, and a linear–quadratic regulator. The controller is intended for the control of linear dynamic systems where the measurements and initial values are corrupted by Gaussian white noise. **S**: LQG controller

**linear-quadratic regulator** A state controller that is the optimal solution to the control problem considering the time interval from the start of observation to infinity. The controlled system is modelled by a set of linear differential equations, whereas the loss function is described by a quadratic function. **S**: LQR

**linear robot** → Cartesian robot

**linear speed** → linear velocity

**linear system** **1**. A system that has linear relations among its input variables, output variables and state variables. It is described by a linear model. **2**. A system of linear equations that define relations among variables from the same set.

**linear valve characteristic** An inherent valve characteristic, which shows a direct proportionality between the flow rate and the valve travel. Change of flow per unit of valve travel remains equal regardless of the valve-disc position. It is used in valves for level control or flow-rate control, especially when a fairly constant pressure drop across the valve is expected.

**linear variable differential transformer** An inductive displacement sensor, which converts linear displacement into an electric signal through the principle of mutual induction. The device consists of an AC-excited primary coil and two secondary coils wound in series but with opposite winding directions. The differential voltage in the secondary coils determines the moved distance. Due to its frictionless construction, it can operate in harsh environments. It is commonly used in, e.g. machine tools, automation, robotics, avionics, computerised manufacturing, and as a primary or secondary transducer in measuring, e.g., weight, pressure, force. **S**: LVDT

**linear velocity** The time derivative of the position of an object. **S**: linear speed

**line contact** The one-dimensional contact with a surface, which has four DOF when there is no friction, and one DOF when friction exists.

**linguistic model** → verbal model

**linguistic variable** A variable, the domain of which is a set of words, rather than a set of numbers, e.g., temperature, expressed as cold, cool, warm or hot.

**link** An element of a stock and flow diagram that denotes a dependency between stocks, flows and dynamic variables. It is usually depicted as a thin arrow, which points from the influencing element towards the influenced element. **S**: action connector, connector, control signal (2), flow of information

**liquid thermometer** A temperature sensor, which consists of a glass bulb attached to a sealed capillary glass tube and a carefully calibrated scale to read off the measured temperature. The bulb is filled with thermally-sensitive medium, mostly mercury or coloured alcohol that expands and rises freely in a tube when its temperature increases. For industrial use, it is mounted in protective housing. It is used in, e.g., industrial applications, medical applications, household applications.

**live zero** A zero-value shift to a nonzero value of the standardised continuous electrical signal or continuous pneumatic signal, e.g., 4 mA in 4-20 mA DC transmission, 20 kPa in 20-100 kPa pneumatic signal transmission. It ensures low sensitivity to induced noise and magnetic interference as occurs when adjacent consumers are activated. Furthermore, it ensures improved signal-noise ratio, fail-safe operation and low sensitivity to voltage drop in the interconnecting wiring. Additional devices can be powered by the current loop itself without the loss of signal.

**LMI** ↔ linear matrix inequality

**load** An element that consumes matter, energy or information, e.g., a motor that uses electrical energy, a machine that processes material, a computer program that requires computing power. **S**: payload

**load capacity** The maximal total mass that can be applied at the end of the robot arm without violating the specifications of the robot.

**load cell** A device, which converts the force caused by tension, compression, pressure or torque, into an electrical signal, e.g., strain gauge load cell, piezoresistive load cell, pneumatic load cell, hydraulic load cell. The measured deformation of the elastically deformed element enables the determination of the force exerted by the load. It is commonly used in, e.g., industry, aerospace, marine, transportation, healthcare.

**load disturbance sensitivity function** The transfer function between the input disturbance and the controlled variable in a closed-loop system. It is used for analysing the influence of the input disturbance on the control variable. **S**: input sensitivity function

**localisation** The process, which enables an autonomous mobile system to establish its own position and orientation within the global coordinate frame. It combines the information from various sensors to compute an estimate of the position and orientation.

**local maximum** The largest element of a set or the largest value of a function over its limited subrange of observation, e.g., a suboptimal solution of an optimisation problem. **S**: relative maximum

**local minimum** The smallest element of a set or the smallest value of a function over its limited subrange of observation, e.g., a suboptimal solution of an optimisation problem. **S**: relative minimum

**logic control** Control, where input and output signals are binary. It is usually implemented with a PLC and used in production systems.

**logic gate** A digital electronic circuit that implements a Boolean function. It is an elementary building block of a digital circuit for, e.g., implementation of combinational control.

**logic model 1**. An abstract symbolic model, which is described by Boolean rules. **2**. A graphic depiction, which presents relationships among the resources, activities, outputs, outcomes and impacts of a program, project or service. If used from the beginning of planning, it helps to guide the design process and its evaluation.

**LonTalk** An open communication network protocol, which is part of the technology platform LonWorks. It is designed to connect devices in a BMS but is also used in, e.g., control system, smart house, smart city, smart grid.

**loop 1**. A closed-shape path of matter, energy or information. **2**. A path in a signal-flow graph or in a block diagram, which originates and terminates in the same node. Along it, no other node is passed more than once.

**loop-transfer recovery 1**. The method for the design of the Kalman filter, which keeps closed-loop system properties obtained with an LQR. **S**: LTR **2**. The method for the design of an LQR, which keeps closed-loop system properties obtained using a Kalman filter. **S**: LTR

**loss function** An objective function, the smaller value of which represents a better result. Therefore, in an optimisation problem, its value is minimised. **S**: cost function

**low-pass filter** A filter, which transmits signals with frequencies lower than the defined cutoff frequency, while signals with higher frequencies are attenuated. The attenuation depends on its order, e.g., 20 dB per decade for first-order and by 40 dB per decade for second-order. It is often used in various signal-processing systems. **S**: high-cut filter

**LQG controller** ↔ linear quadratic Gaussian controller

**LQR** ↔ linear-quadratic regulator

**LTR** ↔ loop-transfer recovery (1, 2)

**Luenberger observer** A linear state observer for estimating the states of a system from the known input signal and output signal. It implements a gain matrix that considers past output-signal measurements and hence influences the estimation-error dynamics. **S**: prediction estimator, prediction observer

**lumped-component model** → lumped-parameter model

**lumped-element model** → lumped-parameter model

**lumped-parameter model** A mathematical model that does not take into account the spatial dimensions of the system elements, e.g., a model represented with ordinary differential equations with time as the independent variable. **S**: lumped-component model, lumped-element model

**lumped-parameter system** A system, the elements of which are concentrated in single points and are not spatially distributed. Therefore, an excitation instantaneously spreads to the whole system, which has only one independent variable, e.g., time. Its mathematical model is described with ordinary differential equations or ordinary difference equations.

**LVDT** ↔ linear variable differential transformer

**Lyapunov's direct method** A method for analysing the stability of equilibrium points of an autonomous system using a suitable Lyapunov function. It is used primarily for stability analysis of nonlinear time-varying systems. **S**: Lyapunov stability criterion, second method of Lyapunov

**Lyapunov's indirect method** A method for analysing the stability of equilibrium points of an autonomous system using linearised approximations in equilibrium points. It is used primarily for stability analysis of nonlinear time-varying systems. **S**: first method of Lyapunov

**Lyapunov function** A continuous scalar function that is positive definite at least locally near the equilibrium point. It is used for a stability analysis, e.g., using Lyapunov's direct method.

**Lyapunov stability** The property of an autonomous system, the solutions of which stay within a limited proximity of an equilibrium point in the state space if the solutions start out near the equilibrium point.

**Lyapunov stability criterion** → Lyapunov's direct method

# Chapter 14
# M

**4–20 mA analogue signal** A standardised electrical analogue signal transmission, the main advantage of which is live zero. It is common in process control to connect process instrumentation with, e.g., controller, PLC, SCADA system.

**machine-human contact** A physical touch between the operator and any part or parts of the robot system. It is the consequence of movements of a part or parts of the robot system.

**machine learning** A data-analysis methodology for automatic mathematical-model development from the relations between data using supervised learning, unsupervised learning or reinforcement learning. The resulting models are, e.g., ANN, fuzzy model, or some less frequently used ones, e.g., spline model, wavelet model, support-vector machine, Gaussian process model.

**machine vision** An engineering technology, which refers to the use of computer vision in a real-life application or process, e.g., automatic inspection, process control, robot guidance. Several technologies are integrated to solve real-world problems.

**magmeter** → electromagnetic flow meter

**magnetic amplifier** A power amplifier, which amplifies the electrical signal utilising a core-saturation principle in a transformer. Alternating current in the secondary winding is controlled by variation of core reluctance due to variation of weak direct current in the primary winding. It is used in, e.g., power supply, measurement system, actuator system, nuclear power plant, arc welder, locomotive.

**magnetic encoder** **1**. A linear encoder, which exploits variations of magnetic flux that is sampled by a corresponding sensing element and converted into an electrical output signal. It can be used in the most demanding and harsh environments for heavy-duty applications requiring robustness, speed, a wide range of operating temperatures, resistance to shock and vibrations, as well as reliability. **2**. A rotary

---

© ZRC SAZU/Research Centre of the Slovenian Academy of Sciences and Arts 2023    123
R. Karba et al., *Terminological Dictionary of Automatic Control, Systems and Robotics*,
Intelligent Systems, Control and Automation: Science and Engineering 104,
https://doi.org/10.1007/978-3-031-35755-8_14

encoder, which exploits variations of magnetic flux that is sampled by a corresponding sensing element and converted into an electrical output signal. It can be used in the most demanding and harsh environments for heavy-duty applications requiring robustness, speed, a wide range of operating temperatures, high-shock and vibrating resistance as well as reliability.

**magnetic flow meter** $\rightarrow$ electromagnetic flow meter

**magnetic thermometer** A temperature sensor, which utilises the measurement of paramagnetic susceptibility of thermally-sensitive material, mostly a nonconducting hydrous rare-earth salt. The material's paramagnetic susceptibility is inversely proportional to its absolute temperature. It is typically used for measuring temperatures below 1 K.

**magnitude** $\rightarrow$ amplitude

**magnitude condition** A point in the $s$-plane, in which the absolute value of an open-loop transfer function equals 1, enabling the determination of the variable parameter value in a certain point of the root locus plot. **S**: magnitude criterion

**magnitude criterion** $\rightarrow$ magnitude condition

**magnitude response** $\rightarrow$ amplitude response

**magnitude scaling** $\rightarrow$ amplitude scaling (1, 2)

**main controller** $\rightarrow$ master controller

**main control loop** $\rightarrow$ master control loop

**Mamdani fuzzy model** A fuzzy model, the consequent part of which is expressed by the membership value appurtenant to a particular output fuzzy set for every individual fuzzy rule. Its output is finally obtained by defuzzification.

**man-in-the-loop simulation** $\rightarrow$ human-in-the-loop simulation

**manipulated signal** $\rightarrow$ controlled signal

**manipulated variable** $\rightarrow$ controlled variable

**manipulative signal** $\rightarrow$ control signal (1)

**manipulative variable** $\rightarrow$ control variable

**man-machine interface** $\rightarrow$ human-machine interface (1, 2)

**manual control** Control, in which a human operator or a supervisor affects the system behaviour with their interventions. It can also be used for automatic-control-systems monitoring and for safety reasons.

**manual load/unload system** A system designed for the direct manual intervention of an operator to interface directly with an industrial robot system, e.g. feed or remove parts or workpieces into and out of a robot cell. It is designed to provide a work area that is free of hazards and to discourage the operator to circumvent the designed safeguarding measures.

**manual-operated valve** → manual valve

**manual valve** A valve, which is manipulated by a human operator, e.g., ball valve, globe valve, butterfly valve, diaphragm valve. It is commonly used in a fluid-transport system or fluid-transport application. **S**: hand-operated valve, hand valve, manual-operated valve

**manufacturing execution system** An information system, which connects, monitors and controls complex manufacturing plants in real time, enabling fact-based operational and strategic decisions. It is connected with the control level, supervisory level and planning level of the automation pyramid. It monitors and synchronises, e.g., product tracking, process scheduling, supply-chain control in real time, and enables paperless manufacturing, elimination of human errors and flexibility. It is used in discrete, batch and continuous manufacturing industries, e.g., semiconductor industry, plastic industry, metal industry, aerospace industry, automotive industry, pharmaceutical industry. **S**: MES

**manufacturing operations and control** → production-control level

**manufacturing resource planning** A computer-based information-management system, which centralises, integrates and processes information for effective decision-making in scheduling, design engineering, inventory managing and cost control of manufacturing systems. It is the predecessor of ERP and is sometimes its module. On the other hand, it is an extension of MRP additionally including machine capacity scheduling, demand forecasting, QA and general accounting. **S**: MRP II

**many-degrees-of-freedom system** The system with several changeable dimensions, parameters or variables, e.g., motion in space, a system with several control inputs, feedback control with controllers in different paths in a signal-flow graph or in a block diagram of the closed-loop system.

**marginally stable system** A linear time-invariant system, the mathematical model of which has at least one pole with the real part equal to 0, or at least one eigenvalue of the system matrix with the real part equal to 0, or at least one root of the characteristic equation with real part equal to 0. All other poles, eigenvalues, or roots have negative real parts.

**marginal stability** The property of a linear, time-invariant system that it is neither stable nor unstable.

**Mason's gain formula** → Mason's rule

126                                                                                  14   M

**Mason's rule** The algorithm, which enables the simplification of a signal-flow graph. As the latter is easily tranformable to the equivalent block diagram the algorithm is usable also for the block diagram simplification. The approach can be computationally supported. **S**: Mason's gain formula

**mass balance** A conservation law, which states that time derivative of stored mass in a mass-storage tank is equal to the difference between the sum of mass flows at the input and the sum of mass flows at the output of the mass-storage tank. **S**: law of conservation of mass, material balance

**mass current** $\rightarrow$ mass flow rate

**mass flow meter** A flow meter, which measures the amount of fluid passing through the pipe, e.g., bulk material mass flow meter, thermal mass flow meter, Coriolis mass flow meter. The measurement is not affected by density, pressure and temperature changes. For fluids with constant density, volumetric-flow-rate measurement is often used.

**mass flow rate** The time derivative of the mass of substance being transferred through a predefined surface or an object, e.g., an orifice. **S**: mass current, mass flux (2)

**mass flux 1**. Mass flow rate divided by the area, through which mass is being transferred. **2**. $\rightarrow$ mass flow rate

**mass spectrograph** $\rightarrow$ mass spectrometer

**mass spectrometer** An analytical instrument, which measures the mass-to-charge ratio of one or more molecules present in a solid, liquid or gaseous sample, enabling the determination of its chemical structure. The vapourised sample is injected into a vacuumed ionisation chamber. The ions are accelerated and deflected by a magnetic field. The deflection depends on the mass and charge of particular ions. The detector counts the ions at different deflections. The resulting mass spectrum shows mass-to-charge ratios of ions against their intensities. It is used in qualitative or quantitative analysis of, e.g., biomolecules, proteins, lipids, peptides, isotopic compositions, respiratory gas, blood samples, urine samples. It is often combined with a gas chromatograph or HPLC. **S**: mass spectrograph, mass spectroscope

**mass spectroscope** $\rightarrow$ mass spectrometer

**master controller** A controller in a cascade control structure, which uses the difference between the adjustable setpoint and the measured controlled variable to create the output, which is the setpoint of the slave controller. **S**: first controller, main controller, primary controller

**master control loop** The exterior loop in a cascade control structure, which contains a master controller, a slave control loop and the controlled process. **S**: main control loop, primary control loop

**material balance** → mass balance

**material requirements planning** A computer-based information-management system, which improves productivity, inventory control and scheduling in manufacturing and process industries, e.g., plans production to meet customers' demand, schedules raw-material as well as components purchase and delivery, supports just-in-time production, assures the lowest level of materials and products in-store. **S:** MRP

**mathematical model** A symbolic model, which is given in the form of an appropriate mathematical structure. It is generally comprehensible and unambiguously interpretable.

**matrix-fraction description** A mathematical model of a MIMO system where a polynomial matrix of numerators of the TFM is divided by the monic least-common denominator of all the elements of the TFM. **S:** MFD

**maximal overshoot** → maximum overshoot

**maximum-likelihood estimation** A method for the optimisation of the parameters of a mathematical model. It searches for the most likely mathematical-model parameters, which fit the model response to the measured system response.

**maximum overshoot** The transient-response specification defining the difference between the maximum peak value and the steady-state value of the response of a proportional underdamped second-order system to a unit-step signal. It is often given in percentage as the maximum peak value minus the steady-state value divided by the steady-state value. **S:** maximal overshoot, peak overshoot, percentage overshoot

**maximum space** Space, which can be reached by the moving parts of the robot as defined by the manufacturer plus the space that can be reached by the robot end-effector and the workpiece.

**McMillan degree** The number of poles in the elements of the Smith-McMillan canonical form of the mathematical model of the MIMO system, described by the corresponding TFM. It determines the order of any minimal realisation of the state-space model.

**McMillan form** → Smith-McMillan canonical form

**MCU** → microcontroller

**measurement range** → measuring range

**measurement system** A subsystem in a control loop, which acquires the quantitative value of a controlled variable. It consists of a sensing element, one or more signal conditioning elements, one or more transducers, a signal processing element, a computation unit and a data-presentation element.

**measurement uncertainty** A metric for the dispersion of the measured values being attributed to a measurand. It can be determined by statistical analysis of the measured results but also taking into account some other sources of mismatch, e.g., calibration, resolution, usage of corresponding probability distributions.

**measuring range** The interval between the minimal value and the maximal value of the measured quantity, in which the defined, agreed or guaranteed error limits are not exceeded. **S**: measurement range, working range

**Mecanum wheel** An omni wheel, which has small rollers usually mounted in such a way that their axis of rotation is at 45° to its plane and its axis of rotation. It enables, e.g., omnidirectional holonomic movement of a 4-wheeled robot. **S**: Ilon wheel

**mechanistic approach** A bottom-up problem-solving technique, which bases on the available data and considers the existing conditions.

**mechatronics** A multidisciplinary branch of engineering, which synergistically integrates electrical components and mechanical components with control systems. It is a unified framework for generating simpler, more economic and reliable systems by combining, e.g., electronics, robotics, computing, communications. It is used in, e.g., control system, manufacturing system, robot system, as well as in automotive engineering, building automation and consumer production.

**medical robot** A robot, which is used in healthcare applications, e.g., surgical robot, rehabilitation robot, robotic exoskeleton, biorobot.

**MelsecNet** A master-slave communication protocol, which allows the connection of up to 64 slave devices in a single loop, e.g., MELSECNET/H, MELSECNET/10. It is used for high-speed delivery of large digital-data volumes in industry.

**membership function** A function, which assigns the membership value of an element to a particular fuzzy set. The membership value is within the interval between 0 and 1. **S**: blending function, interpolation function, validity function

**membership value** A quantitative measure of the membership of an element to a particular fuzzy set. It is usually defined by the set of membership functions. **S**: degree of fulfilment

**membrane pressure sensor** → diaphragm pressure sensor

**membrane pump** → diaphragm pump

**membrane valve** → diaphragm valve

**MEMS** ↔ microelectromechanical system

**MEMS pressure sensor** An absolute pressure sensor, which integrates a sensing element, most commonly deformable diaphragm, and a digital conditioning chip using capacitive or piezoresistive signal transduction. Compensation concerning drift, sensitivity, linearity is also included. It measures the pressure of noncorrosive pure gases.

**mental model** A subjective abstract representation of the external reality, which uses human experiences as the basis for the reasoning and decision making. Since it is intuitive and heuristic it may make the communication ambigous due to the imprecise formulation.

**MES** ↔ manufacturing execution system

**meta-parameter** → hyperparameter (1, 2)

**MFD** ↔ matrix-fraction description

**Michelson interferometer** → laser interferometer

**microactuator** A submicrometre-size actuator to millimetre-size actuator, which is produced using silicon-chip technologies or a precise version of conventional machining, e.g., laser machining. It can be used as a component of MEMS as well as for the generation of micromotion using, e.g., electrostatic principle, electromagnetic principle, piezoelectric principle, thermal principle, and for the generation of microfluid in, e.g., implantable drug delivery system.

**microbotics** → microrobotics

**microcomputer controller** A stand-alone, single-loop or multi-loop controller, which consists of a microprocessor, memory, input/output units and a front-panel with a display and a keypad or a touchscreen for monitoring and parametrisation.

**microcontroller** A single-chip microcomputer, which is designed to perform specific tasks of an embedded system. It consists of a processor, memory and peripherals, e.g., timer, counter, interrupt control, input/output port, A/D converter, D/A converter. It is of small size, inexpensive and is used in, e.g., robotics, communication systems, biomedical instrumentation, low-cost wearables, automotive industry, intelligent housing. **S:** $\mu$C, MCU, uC

**microelectromechanical system** A structure of mechanical and electrical components between 1 and 100 micrometres in size, which often includes sensors, actuators as well as computer and communication facilities, e.g., accelerometer, micro-optical component, microfluidic component, medical device. **S:** MEMS

**microrobotics** Robotic manipulation of objects with dimensions in the range from micrometre to millimetre as well as the design and fabrication of autonomous robotic agents that fall within this size range. **S:** microbotics

**microrobot system** A robot system, which is composed of micromanipulators, micromachines, and man-machine interfaces, e.g. haptic interface.

**microsensor** A submicrometre-size sensor to millimetre-size sensor, which is produced using silicon-chip technologies. It can be used as a component of MEMS as well as in, e.g., process industry, automotive industry, food industry, environment applications, medical applications, consumer electronics.

**microwave density meter** A density meter, which determines the consistency of fluids flowing through pipes by monitoring the propagation of electromagnetic waves in the medium. Either the phase shift between the original wave and the one that passes through the measured fluid, or the attenuation of the transmitted wave in the medium enables the calculation of the measured density. It is used in, e.g., cement industry, pulp industry, food industry, building construction, wastewater treatment.

**MIMO system** ↔ multiple-input multiple-output system

**minimal realisation** State-space description of the controllable and observable system with the smallest possible set of state variables. However, the description is not unique. The pole-zero cancellations are required in the transfer-function notation.

**minimum-order observer** A system for the estimation of the values of nonmeasured system states only. It gathers information from measured signals, usually the input signal and the output signal. The measured system states are obtained directly. **S**: reduced-order observer

**MISO system** ↔ multiple-input single-output system

**mixing valve** A three-way valve, which enables the combination of two inlet streams of fluid into one outgoing stream, e.g., blending hot water with cold water ensuring a constant outlet temperature. It is used in, e.g., residential application, commercial application.

**mixproof valve** An on-off valve, which enables two fluids to flow through it without risking cross-contamination. It has two independently controlled valve stems. Therefore, it can operate with several incompatible fluids with leakage-free switching. The valve has a leakage chamber, which can be independently flushed and drained, e.g., with the cleaning fluid. This prevents intermixing of the media. It is often used in pipeline processes when an aseptic operation is required, e. g., in the clean-in-place system. **S**: double-seat valve (2)

**MNA method** ↔ multivariable Nyquist-array method

**mobile mapping** A process of collecting geospatial data using a mobile vehicle or an autonomous mobile system. A map of the environment is created by implementing various remote sensing systems. The map can be used for decisions about corresponding movements.

**mobile robot** A mobile system, which is capable to move in its environment. It uses, e.g., wheels, tracks, legs, propellers, jet engines to move on the ground, in the water or in the air.

**mobile robotics** A subfield of robotics that studies, designs and implements moving mechanisms utilising artificial intelligence methods.

**mobile servant robot** A personal care robot, which is capable of motion necessary to perform serving tasks in interaction with humans, e.g., handling objects, exchanging information.

**mobile system** A system, which is not attached to the environment and can freely move around, e.g., ground mobile system, aerial mobile system, water mobile system, underwater mobile system. It is often energy-independent.

**mockup** → prototype (1, 2)

**modal canonical form** → diagonal canonical form

**Modbus** A master-slave serial industrial Ethernet, which is open, vendor-neutral and enables the connection among intelligent devices for industrial automation systems and BMS as well as for messaging in intranet or internet environments using TCP/IP protocol, e.g., Modbus RTU, Modbus TCP.

**model** A simplified, physical or abstract, representation of the real system, emphasising the aspects that are important according to the specified modelling goal. The reality is transformed into a comprehensible and usable form on a different media, e.g., paper, computer memory.

**model accreditation** An official recognition that a model has sufficient model credibility to be used for the specified purpose in the particular problem domain.

**model-builder's risk** A situation in the model-development procedure, where the modeller does not trust the designed model due to inadequate validation and unjustifiably modifies it further.

**model credibility** A measure of the user's or the modeller's confidence that modelling results are acceptable for the prespecified purpose in the particular problem domain.

**model decomposition** A process of breaking down a mathematical model of a complex system. It starts from general characteristics of the system towards more specific ones. For every level of abstraction, the corresponding submodels are developed. The solutions of submodels lead also to the solution of the original model.

**model deduction** A procedure, which starts with a simple model and gradually upgrades it by including more details. It ends when reaching the appropriate complexity that meets the required accuracy specification.

**model-driven development** A software engineering approach that uses a model to create a product. The gap between the specification and the implementation may be bridged by an automated transformation.

**model-fit criterion** A statistical method that enables the evaluation of similarity between the model response and the measured output data or the prescribed curve as well as a comparison of different models. **S**: goodness-of-fit criterion

**model fitting** → curve fitting

**model-identification adaptive control** → self-tuning control

**model library** A collection of well-tested and numerically robust mathematical models from a certain domain, co-created by domain experts. Its components must be designed in a predefined form, which makes them usable in diverse applications.

**modelling** Various complex activities that make up the process of model development aiming to improve, e.g., the understanding of the functioning mechanisms of the modelled system, the prediction of its behaviour, the design and the evaluation of control systems, the estimation of the unmeasurable system states, the optimisation of the system behaviour, the development of simulators, as well as to enable the sensitivity analysis and the fault diagnosis.

**modelling environment** → simulation environment

**model management** A planning system for documentation and maintenance of mathematical models, which enables well-maintained libraries of reusable models.

**model-order reduction** A procedure, which lowers the complexity of a mathematical model in numerical simulation by reducing the dimension of the model. **S**: MOR

**model-prediction uncertainty** A range of differences from the most likely predicted value, which is the consequence of possible input uncertainties, parameter uncertainties and structural uncertainties as well as of the bias of the mathematical model. Several methods enable the evaluation of modelling results' reliability.

**model predictive control** Control of a dynamic system, which implements a mathematical model of the controlled system to simulate its future behaviour with regard to the control variable. The implemented value of the control variable in a particular time step is usually established using optimisation. **S**: moving-horizon control, MPC, receding-horizon control

**model reduction** A procedure that enables the determination of a mathematical model, which is as simple as possible, but still satisfactorily approximates the behaviour of the originally developed complex model. **S**: model simplification

**model-reference adaptive control** An adaptive control that achieves the best fit between the system output and the output of a mathematical model, which expresses the desired output of the system.

**model-reference adaptive controller** An adaptive controller, which causes the controlled closed-loop system to imitate the behaviour of the predefined mathematical model expressing the desired output of a system. Online identification of system parameters is not needed.

**model simplification** → model reduction

**model-structure optimisation** A method for the determination of the structure of a mathematical model, which considers the selected objective function and the preset constraints.

**model-user's risk** A situation in the model-development procedure, where the user unjustifiably applies the model for a purpose for which it has not been developed. Such abuse of the model can lead to wrong interpretation of the modelling results.

**model validation** A procedure, which evaluates whether the model behaviour is satisfactorily similar to the behaviour of the modelled system. In such a case the model can be regarded as an adequate representation of reality.

**mode of operation** The operation of a system, which is defined by load rate or one of the attainable ways of functioning. **S**: operational regime

**modified Routh-Hurwitz stability criterion** A Routh-Hurwitz stability criterion, which is adapted to discrete-time systems, where an adjusted variable enables the determination of equivalent continuous-time poles. The stability of the system depends on the number of poles outside the unit circle, which is determined from the number of equivalent continuous-time poles in the right $s$-halfplane of the adjusted variable.

**modified Routh stability criterion** → modified Routh-Hurwitz stability criterion

**modular design** A reduction or decomposition of a design process to corresponding independent design subproblems, the sum of which represents the complete design procedure. **S**: componental design

**modular PLC** A PLC, which consists of an expandable input and output units, a central processing unit and power supply, housed in a metal framework with multiple slots. Additional modules can be added, e.g., outputs, memory, A/D converter, D/A converter. This enables easy configuration changes, customisation and ability to control complex processes.

**modular robot** A robot, which consists of independent building elements, e.g., robot links, robot joints, actuators, combined into diverse kinematic structures. Each can have some level of independent control. Therefore, the flexibility, robustness and performance of such structure are increased.

**modular robot system** A robot system, which consists of several units with a few DOF, equipped with connecting mechanisms, e.g., a chain unit, a legged unit, a wheeled unit. It supports the creation of various robot configurations that use the corresponding collaborative strategies to meet the requirements according to the possible applications.

**moisture meter** → moisture sensor

**moisture sensor** A humidity sensor, which measures the water content in solid materials directly or by measuring the humidity of the surrounding gas, e.g, gravimetric moisture sensor, resistive moisture sensor, capacitive moisture sensor, microwave moisture sensor, infrared wave moisture sensor, radio-frequency wave moisture sensor. It can measure humidity in a destructive or nondestructive manner. **S**: moisture meter

**momentum balance** A conservation law, which states that the total momentum in a closed system is constant. The total momentum is the sum of partial momenta. **S**: momentum conservation

**momentum conservation** → momentum balance

**momentum wheel** A flywheel, which compensates for short-lasting outside disturbances with constant rotation, usually in a spacecraft. Therefore, it replaces thrusters or external applications of torque.

**Monte Carlo simulation** A statistical simulation method, which uses probability distribution to model a random input variable. The distribution of possible outcomes is generated iteratively, using different sets of random values. It is used in, e.g., research, engineering, artificial intelligence, manufacturing, project management, finance, supply chain, computational biology.

**MOR** ↔ model-order reduction

**motion detector** → motion sensor

**motion planning** A computational problem in robotics encompassing path planning and determination how to move. It bases on velocity, time, and kinematics.

**motion sensor** A sensor, which notices moving objects, mainly people, e.g., passive infrared sensor, ultrasonic sensor, digital camera. It is commonly used in indoor and outdoor systems, e.g., home security system, lighting system, energy-efficiency system. **S**: motion detector

**moving-horizon control** → model predictive control

**MPC** ↔ model predictive control

**MRP** ↔ material requirements planning

**MRP II** ↔ manufacturing resource planning

**multi-agent system** A set of autonomous or partly autonomous agents, which cooperate, compete and share knowledge among themselves as well as with their environment to fulfil a common or an individual goal. It can tackle problems that are difficult or impossible to be solved by a single agent or a monolithic system. It is used in, e.g., distributed robotics, logistics, simulation, grid computing, robot soccer, computer games, economics.

**multicriteria optimisation** → multi-objective optimisation

**multi-domain modelling** Physical modelling of a complex system, representing a type of virtual prototyping. It results in an acausal model, which includes components from various problem domains, e.g., mechanics, electronics, hydraulics, pneumatics, optics. It is often implemented in, e.g., mechatronic application, automotive application, aerospace application, robotic application.

**multi-faceted modelling** Modelling, in which the system is described with multiple mathematical models of different complexity and for different purposes. It enables solving various engineering problems or solving one engineering problem in different ways.

**multifingered hand** A robot gripper with multiple finger-shaped mechanisms, which are all connected to the same common palm-shaped robot link.

**multilayer perceptron** A feedforward neural network composed of several perceptrons that are organised in layers, which are connected in a cascade.

**multilegged robot** A robot with a high DOF, which is well suited for noneven, rough and unstructured terrain. It imitates the locomotion of quadrupeds, six-legged insects, eight-legged spiders or organisms with a large number of legs like centipedes.

**multi-level control** The implementation of a control algorithm, the result of which is a control variable with a value from a finite set of discrete values. **S**: multi-step control

**multi-level controller** A controller, the output of which has one value from a finite set of discrete values, e.g., on-off controller, three-level controller. **S**: multi-step controller

**multi-objective optimisation** An optimisation method, which considers several goals simultaneously. Its objective function consists of weighted elements. They may be conflicting so that none of the objective subfunctions can be improved in value without degrading some of the other objective function values. **S**: multicriteria optimisation, multi-objective programming, Pareto optimisation, vector optimisation

**multi-objective programming** → multi-objective optimisation

**multiple-input multiple-output system** A system with several inputs and outputs. Its structure is defined by direct input-output connections as well as by cross-couplings. Therefore, each input may affect each output. Furthermore, each output may affect also the other outputs. **S**: MIMO system, multivariable system

**multiple-input single-output system** A dynamic system, which has several inputs and one output. Such structure enables the development of observable canonical form and thus facilitates state observer design. **S**: MISO system

**multiple-model system** A globally valid mathematical model of a nonlinear system comprised of a set of locally valid mathematical models, e.g., Takagi-Sugeno fuzzy model, piecewise affine system.

**multi-robot system** A system, where two or more simple robots cooperate and interact with each other to perform complex tasks, which are difficult or impossible to be accomplished by a single more complex robot.

**multistage amplifier** An amplifier, which consists of two or more same-type or different-type amplifiers that are used in series to improve its performance. Its stages can be either electrical, pneumatic or hydraulic.

**multi-step control** → multi-level control

**multi-step controller** → multi-level controller

**multi-step integration method** A numerical integration method, which requires the value of the integral in several previous steps to calculate the value of the integral in the current step, e.g., trapezoidal method, Adams-Basforth method, Adams-Moulton method.

**multivariable control** Control, for which the design goals require the use of a controller with the character of a MIMO system.

**multivariable Nyquist-array method** A MIMO-control-design method, i.e., INA method, DNA method, which is based on the fact that the complete decoupling of a MIMO system is often not necessary to achieve the design requirements. Therefore, the cross-couplings are only sufficiently attenuated according to the corresponding graphical criteria. **S**: MNA method

**multivariable system** → multiple-input multiple-output system

# Chapter 15
# N

**nanoactuator** A molecular-scale actuator, which is produced using technologies that can manipulate atoms and molecules.

**nanoelectromechanical system** A structure of mechanical and electrical components below 1 micrometre in size, which often includes sensors, actuators as well as computer and communication facilities, e.g., accelerometer, nano-fluidic module, sensor for detecting chemical substances in the air. It is the next step in miniaturisation from MEMS and can provide the possibility of observing quantum effects. **S:** NEMS

**nanorobotics** Robotic manipulation of objects with dimensions smaller than a micrometre as well as the design and fabrication of autonomous robotic agents that fall within this size range.

**nanosensor** A molecular-scale sensor, which is produced using technologies that can manipulate atoms and molecules. It can be used as a component of MEMS or NEMS as well as in, e.g., defence and military applications, food and environment applications, healthcare.

**natural frequency** The frequency at which the output response of a linear dynamic system tends to oscillate in the absence of damping or any excitation. **S:** eigenfrequency, undamped frequency

**natural response** A time response of a dynamic system, which is excited solely with initial conditions. In case there are any auxiliary inputs, they are all 0. It is the homogeneous solution of the differential equation that describes the mathematical model of the observed system. **S:** characteristic response, free response, unforced response

**nature-inspired robotics** → biologically-inspired robotics

**nautical angles** → Tait-Bryan angles

© ZRC SAZU/Research Centre of the Slovenian Academy of Sciences and Arts 2023
R. Karba et al., *Terminological Dictionary of Automatic Control, Systems and Robotics*,
Intelligent Systems, Control and Automation: Science and Engineering 104,
https://doi.org/10.1007/978-3-031-35755-8_15

**navigation** The process of planning and directing a mobile system along a route to move safely from the current location to the designated location without colliding with other objects. It consists of localisation, path planning and mobile mapping.

**NC contact** ↔ normally-closed contact (1, 2)

**needle valve** A valve, which accurately controls mostly low flow rates of clean gasses or fluids using a valve disc with a spiked and tapered end. It is used in metering applications of, e.g., steam, gas, oil, water, nonviscous fluid as well as in, e.g., combustion control system, low-pressure hydraulic system, chemical processing.

**negative feedback** A feedback, in which the setpoint value and the output value are subtracted, resulting in control error. It is in the opposite phase with the input, tending to push the output towards the desired value. It has a stabilising effect and reduces the impact of disturbances and noise. It appears in, e.g., control systems, electronic engineering, mechanical engineering, environmental applications, chemistry, biology, economy, sociology, psychology. S: balancing feedback, degenerative feedback

**NEMS** ↔ nanoelectromechanical system

**nephelometer** A photometer, which measures scattered light from a cuvette containing suspended particles in a solution, in a gaseous liquid or in a transparent solid. It consists of a visible-light source or a laser-light source, a cuvette with a sample and a detector, mostly a photomultiplier tube that is placed at a certain angle from the incident light. The amount of scattered light depends on the size and number of particles in the suspension. It is calibrated to a known particulate considering also the corresponding compensations. It is used in, e.g., water pollution monitoring, air pollution monitoring, climate monitoring, biological contaminants monitoring, inorganic analysis.

**nested method** **1.** A method, which enables the development of a simulation scheme for a mathematical model described by a differential equation that includes time-derivatives of the input variable, or by a transfer function with the numerator of order higher than 0. By nesting the terms according to the order of time-derivatives, the differential equation is rearranged, so that it results in the observable canonical form of the simulation scheme. **2.** A method, which enables the development of a simulation scheme for a mathematical model described by a difference equation that includes delayed samples of the input variable, or by a transfer function with the numerator of order higher than 0. By nesting the terms according to the order of delays, the difference equation is rearranged, so that it results in the observable canonical form of the simulation scheme.

**neuro-fuzzy model** A mathematical model with the structure and the parameters of a fuzzy model, which is trained like an ANN to determine the values of its parameters.

**Newton method** → Newton-Raphson method (1, 2)

**Newton-Raphson method** **1**. A root-finding algorithm for solving nonlinear algebraic equations, which produces a successively better tangent approximation to the roots of continuously differentiable function from a function itself and its derivative, starting in an initial guess. **S**: Newton method **2**. A root-finding algorithm for solving optimisation problems, which determines the extremum of a twice-differentiable function as a root of its derivative. Successively better tangent approximation to the roots is obtained from the first and the second derivative of the function. **S**: Newton method

**Nichols chart** A diagram showing the frequency response of a system, so that the $x$-axis represents the phase angle, and the $y$-axis represents the amplitude in decibels. Each point of the diagram represents the frequency response at a particular frequency. The frequency is thus the parameter, ranging from 0 to infinity. **S**: Nichols diagram, Nichols plot

**Nichols diagram** $\rightarrow$ Nichols chart

**Nichols plot** $\rightarrow$ Nichols chart

**NO contact** $\leftrightarrow$ normally-open contact (1, 2)

**node** **1**. An equilibrium point of a second-order dynamic system in the phase plane, which is the source or sink of the arbitrarily-shaped trajectories. **2**. A point in a signal-flow graph, which represents either a variable or a signal, i.e., input node with outgoing branches only, output node with incoming branches only, mixed node with incoming and outgoing branches.

**noise** **1**. Random test signal with specific statistical properties. **S**: stochastic signal **2**. Unwanted random changes in the measured signal, e.g., thermal noise, noise due to vibrations.

**noise level** A quantity of noise that is usually expressed as the mean square of the noise signal, the power spectral density of the noise signal or the probability distribution of noise as a random variable.

**noise sensitivity function** The transfer function between the negative output disturbance and the control variable in a closed-loop system. It is used for analysing the influence of the output disturbance on the control variable. **S**: output sensitivity function

**nominal performance** A closed-loop system behaviour, which is evaluated by the aid of sensitivity function and complementary sensitivity function.

**non-interacting system** $\rightarrow$ decoupled system

**non-interactive system** $\rightarrow$ decoupled system

**nonlinear model** A model described with nonlinear mathematical structures, e.g., with nonlinear differential equations. The law of superposition can not be applied.

**nonlinear system** 1. A system that has nonlinear relations among its input variables, output variables and state variables. Consequently, the change of its output is not proportional to the change of its input. It is described by a nonlinear model. **2**. A system of nonlinear equations that define relations among variables from the same set.

**nonmathematical model** An abstract symbolic model, which describes a system verbally, graphically or schematically. It is often ambiguous.

**nonminimum-phase system** A linear system, the continuous-time transfer function of which has zeros in the right-half of the $s$-plane, or the discrete-time transfer function of which has zeros outside the unit circle in the $z$-plane. Its amplitude response is equal to the corresponding minimum-phase system with the mirrored zeros in the left half of the $s$-plane for the continuous-time transfer function or the zeros mapped into the unit circle in the $z$-plane for the discrete-time transfer function. However, its phase response is different, and therefore its step response has an overshoot in the direction opposite to the new steady state.

**nonparametric identification method** The identification method that results in tables of values or in a graphical representation, e.g., Fourier analysis, frequency response analysis, correlation analysis, spectral analysis.

**nonparametric model** A mathematical model with implicitly included parameters, e.g., input-output behaviour presented with tables of values or with a graphical representation, such as time response or frequency response.

**nonreturn valve**  $\rightarrow$ check valve

**non-SI unit**  $\rightarrow$ engineering unit

**nonsquare MIMO system** A MIMO system with an unequal number of inputs and outputs described with a nonsquare TFM.

**normal form**  $\rightarrow$ canonical form

**normalisation** The adjustment of the data values measured on different scales to a common scale enabling the comparison of the normalised values of various data sets.

**normally-closed contact** 1. A contact in electrical switches, relays, connectors, in which the signal path is uninterrupted in the default state, which is the nonpowered state. When powered, the signal path becomes interrupted. **S**: NC contact **2**. An element of a ladder diagram, the signal path of which is uninterrupted when logic input is 0, and interrupted when logic input is 1. **S**: NC contact

**normally-open contact** 1. A contact in electrical switch, relay, connector, in which the signal path is interrupted in the default state, which is the nonpowered state. When powered, the signal path becomes uninterrupted. **S**: NO contact **2**. An element of a ladder diagram, the signal path of which is interrupted when logic input is 0, and uninterrupted when logic input is 1. **S**: NO contact

**nuclear densitometer** → nuclear density gauge

**nuclear density gauge** A density meter, which operates on the principle of gamma-ray absorption that increases with the density of the measured material. It consists of a gamma-ray source and an appropriate detector, often a Geiger-Müller tube. Gamma rays lose energy through interaction with the material or they are scattered away from the detector, the count of which is proportional to the measured density. It is used for compactness measurements of solid materials or liquids in a pipe. **S:** nuclear densitometer

**nuclear level sensor** → radiation level sensor

**numerical approximation error** → truncation error

**numerical integration method** Approximative computation of an integral using numerical techniques. It enables solving differential equations with successive integration.

**Nyquist contour** A closed semicircular contour with an infinitely large radius, going clockwise and encompassing the whole right $s$-halfplane. It excludes the singularities of the open-loop transfer function on the imaginary axis by avoiding them along semicircular diversions with infinitely small radii.

**Nyquist diagram** → Nyquist plot

**Nyquist frequency** The highest allowable frequency in the spectral density of the original frequency-limited continuous-time signal being sampled that enables theoretically perfect reconstruction of the original signal from its samples. It is equal to one half of the sampling rate.

**Nyquist plot** A closed contour in the complex plane, which is the result of mapping the Nyquist contour via the open-loop transfer function. It is used for stability analysis using the Nyquist stability criterion. **S:** Nyquist diagram

**Nyquist rate** The lowest allowable sampling rate of a frequency-limited continuous-time signal that enables theoretically perfect reconstruction of the original signal from its samples. It is equal to double the highest frequency in the spectral density of the original signal.

**Nyquist-Shannon theorem** → sampling theorem

**Nyquist stability criterion** A graphical stability-analysis method for linear time-invariant closed-loop systems. It determines the stability from the difference between the number of the Nyquist plot encirclements of the $-1+0i$ point, and the number of nonstable open-loop poles. The system is stable when the difference is nonnegative. **S:** Strecker stability criterion

# Chapter 16
# O

**objective function** A mathematical function, which represents a quantitative performance measure for a certain algorithm, e.g., optimisation. In the case of maximisation, a bigger value represents a better result, whereas in the case of minimisation, a smaller value represents a better result. **S**: criterion function

**object linking and embedding for process control protocol** $\rightarrow$ open platform communications classic

**object-oriented modelling** A modular modelling-based programming paradigm, which provides a clear model-structure description by connecting predefined objects. Features, such as acausal modelling with equations, simple evolution of models, easy creation of components, multidomain modelling capability, and reuse of components, where detailed knowledge of their internal structure is not required, represent an advantage in the development of complex mathematical models.

**observability** The property of a system, the initial state of which can be determined in finite time from the known system output.

**observable canonical form** A canonical form of a state-space model, which ensures the observability of the system. It can be obtained, e.g., with the transformation of the transfer function into the simulation scheme using the partitioned method. The description enables a direct determination of the characteristic polynomial and the transfer function of the system from the rightmost column of the obtained system matrix. **S**: observer canonical form

**observer canonical form** $\rightarrow$ observable canonical form

**obstacle avoidance** The strategy of a motion of a mobile robot that enables the design of a modified path plan when an obstacle is detected. The mobile robot can determine a collision-free trajectory using various sensors, e.g., ultrasonic sensor, infrared sensor, laser distance sensor. Often, machine vision and a map of the surrounding area are also used. **S**: collision avoidance (1)

© ZRC SAZU/Research Centre of the Slovenian Academy of Sciences and Arts 2023    143
R. Karba et al., *Terminological Dictionary of Automatic Control, Systems and Robotics*,
Intelligent Systems, Control and Automation: Science and Engineering 104,
https://doi.org/10.1007/978-3-031-35755-8_16

**Occam's razor** The principle used also in modelling, which states that the model should be kept as simple as possible, with respect to the modelling goal. **S**: law of parsimony, principle of parsimony

**octave** A unit for measuring frequency ratio of two, usually between two frequencies on a logarithmic scale, defining the frequency range between a particular frequency and its twofold value. It is used for expressing slopes of asymptotes in the Bode magnitude plot and Bode phase plot of the asymptotic Bode plot.

**odometer** An electronic, mechanical or electromechanical device, which measures the distance travelled by a mobile system.

**odometry** A measurement of distance, which uses information from motion sensors. It estimates the route travelled by a mobile system, e.g., from the measurement of the angular displacement of the mobile robot wheels or by analysing a video stream.

**OE model** ↔ output-error model

**offline method** A method where the complete dataset must be available before its use. The possible new data, obtained during implementation, are not taken into account.

**offset** The constant difference between the instrument reading and the measured value across the entire measuring range.

**offset error** → steady-state error

**OLE for process control protocol** → open platform communications classic

**omnidirectional mobile robot** A highly manoeuvrable mobile robot, which can move instantaneously in any direction regardless of its current orientation, as well as reorient itself at the spot. It can be designed with conventional wheels or with special wheels, e.g., omni wheels, universal wheels, ball wheels.

**omnidirectional wheel** → omni wheel

**omni wheel** A wheel, which provides a minimum amount of friction sideways, allowing it to move in any direction. It has small rollers along its edges. The axes of the rollers are often perpendicular to its axis of rotation. **S**: omnidirectional wheel

**one-degree-of-freedom system** → single-degree-of-freedom system

**one-step-ahead prediction** The prediction of a regression model's output value based on the input values from the previous time step.

**one-step integration method** → single-step integration method

**one-way valve** → check valve

**online method** A method, which processes constantly streaming data in real time.

**on-off actuator** An actuator with two discrete output states, e.g., shut-off valve, pneumatic relay, solenoid. **S**: discrete actuator

**on-off control** The implementation of a control algorithm, the result of which is a control variable with a value from a set of two discrete values, which are usually 0 and 1. **S**: bang-bang control, two-level control, two-step control

**on-off controller** A controller, the output of which has one of the two possible values, which are usually 0 and 1. **S**: bang-bang controller, two-level controller, two-step controller

**on-off sensor** → discrete sensor

**on-off valve** A valve, which either allows unimpeded flow or completely stops the flow, thus isolating the neighbouring pneumatic or hydraulic subsystems, e.g., ball valve, butterfly valve, gate valve, diaphragm valve. It is the fluid equivalent of the electrical switch. It is utilised in residential, commercial and industrial applications, e.g., at the event of a safety risk, at equipment failure, at carrying out maintenance. **S**: shut-off valve

**opacimeter** → turbidimeter

**op-amp** → operational amplifier

**OPC** ↔ open platform communications

**OPC classic** ↔ open platform communications classic

**OPC UA** ↔ open platform communications unified architecture

**open kinematic chain** A kinematic chain, the elements of which are connected in series.

**open-loop control** A simple control strategy, which uses pre-known responses of the control system to the input signal. The controlled variable does not influence the control variable. Consequently, there are no self-correcting reactions to disturbances and changes in the system.

**open-loop observer** A model of the system, which runs in parallel to the system and is used for estimating the values of system states by considering the known input signal. The error of the estimate converges to 0 if the system is asymptotically stable.

**open-loop system** A system, which does not contain any feedback path, or the latter is interrupted.

**open-loop transfer function** A transfer function, which is the product of all transfer functions along the forward path and along the feedback path, connected in series.

**open platform communications** A platform-independent interoperability standard for data exchange that comprises a series of specifications, e.g., OPC classic, OPC UA. It defines the interface between clients and servers, as well as among servers, including access to real-time data, monitoring of alarms and events, access to historical data. It is used in industrial automation. **S**: OPC

**open platform communications classic** An OPC that allows programs developed for the Microsoft Windows operating system to communicate with industrial hardware devices, enabling the connection of HMI or SCADA systems with controllers and PLCs. It uses the COM/DCOM model for the exchange of data between software components. It provides separate specification definitions for accessing process data, for alarms and events, as well as for historical data, e. g., OPC DA, OPC AE, OPC HDA. **S**: object linking and embedding for process control protocol, OLE for process control protocol, OPC classic

**open platform communications unified architecture** An OPC that enhances the functionality of OPC classic specifications. The OPC data models can be implemented on Microsoft and non-Microsoft systems, including embedded devices. Furthermore, it enables users to access OPC servers through firewalls in a secure manner. **S**: OPC UA

**operating point** **1**. A point on the static characteristic of a system, in the neighbourhood of which the operation of the system is intended. It is determined with the values of system input and system output or with the values of system input and system states. **2**. The origin of a coordinate system, which is placed according to the definition of the deviation model of a process. **3**. A point on the nonlinear static characteristic of a system, in the neighbourhood of which the linear approximation is used.

**operating range** An interval between the specified limits of the controlled variable for the normal operating conditions.

**operating space** The subspace within the restricted space, which is used while performing all motions commanded by the task program of a robot.

**operational amplifier** A DC-voltage amplifier, which has high gain, high input impedance, low output impedance, large bandwidth, and low input offset voltage. It is implemented as an integrated circuit and enables amplification, transformation, conversion or combination of signals and includes different compensations as well as impedance adjustments. It often performs mathematical operations on the input voltage. **S**: op-amp

**operational regime** $\rightarrow$ mode of operation

**operational research** $\rightarrow$ operations research

**operations management** A multidisciplinary field, which utilises resources, e.g., staff, materials, equipment, technology, to manage and to optimise strategic as well as day-to-day production of goods and services. It determines the necessary improvements for better efficiency. It converts inputs, e.g., raw materials, labour, energy, to outputs, e.g., products, services, using supply chain management and logistics to achieve the required target, e.g., quality, price, flexibility, availability, minimal waste.

**operations research** The application of advanced analytical methods for the optimisation-based determination of the quantitative solution of a problem.

**operator-interface terminal** → operator panel

**operator panel** An interface device, which enables simple communication between the operator and the control system in an industrial environment. It is used to monitor, perform diagnostics, adjust system settings and display information graphically. It enables a flexible configuration, installation and upgrading with a keypad or a combination of a keypad and a touch screen. **S**: HMI terminal, operator-interface terminal, operator terminal

**operator terminal** → operator panel

**optical encoder** **1**. A linear encoder, which uses a photosensitive element to identify the positional changes of the measured object. The light passing through or reflecting from a patterned plate generates the corresponding binary output signal by photoelectric conversion. It is used in, e.g., CNC milling, robotic-arm end-effector, surgical robotics, DNA sequencing. **2**. A rotary encoder, which uses a photosensitive element to identify the orientational changes of the measured object. The light passing through or reflecting from a patterned plate generates the corresponding binary output signal by photoelectric conversion. It is used in, e.g., motor-speed measurement, CNC milling, robotic-arm end-effector, surgical robotics.

**optical level sensor** A level sensor, which consists of a light source leading infrared light in a transparent conic prism, and a phototransistor for detecting the reflected light. When the prism is in contact with the measured liquid, the light is refracted causing less energy to reach the detector. The portion of the returned light indicates the presence or absence of liquid. It is used for point-level detection and as a level switch for liquids.

**optical pressure sensor** An absolute pressure sensor, which detects deflections of a dedicated primary element caused by the measured pressure. The primary element is often a diaphragm, on which an opaque vane is attached that changes the light intensity transmitted between the light source and the light detector. It can also be a reflective diaphragm, which reflects the light, conducted by an optical fibre from the light source and uses another optical fibre to conduct the reflected light to the light detector. Alternatively, the bending of the diaphragm is measured by a fibre optic sensor that uses interferometry to measure path length and phase shift of light.

**optical proximity sensor** → photoelectric proximity sensor

**optical pyrometer** A pyrometer, which bases on the principle of matching the brightness of a filament placed in housing to the brightness of the measured object using a lens that collects the emitted visible light. When both brightnesses are equal, the temperatures of the filament and of the object are the same. The temperature of the filament is calculated from the electrical current heating it. The matching can be performed manually or automatically. It is highly accurate and often portable. It is used in, e.g., furnace control, semiconductor manufacturing, glass manufacturing, medical device.

**optical tachometer** A noncontact tachometer, which emits a light beam that is reflected from a piece of reflecting tape attached to the rotating object. Its angular velocity is calculated from a frequency of reflections.

**optimal control** A control algorithm that tends to find the best possible solution over a period of time in controlling a dynamic system using optimisation. The prescribed objective function is minimised or maximised considering prespecified constraints. It is commonly used in e.g., engineering, science, operations research.

**optimisation** A branch of applied mathematics, which searches for the best solution to a problem. It selects the best possible element from a set of available alternatives according to a given criterion, often searching for the extremum of a function within the prescribed limits. It is used in, e.g., control engineering, manufacturing, transportation, finance, economics, marketing.

**optimisation method** A numerical algorithm, which seeks the optimum of an objective function, e.g., Hooke-Jeeves method, Nelder-Mead method, Davidon-Fletcher-Powell method, Fletcher-Reeves method.

**optocoupler** A single-package semiconductor device, which transfers a signal between two galvanically isolated circuits by using light, with, e.g., light-emitting diode, infrared emitting diode, laser diode, phototransistor, photodiode. It can transfer digital signals as well as analogue signals in, e.g., control, monitoring, communication. **S**: optoisolator, photocoupler

**optoisolator** → optocoupler

**orientation** Three rotational DOF describing the direction of an object in space.

**orifice pass area** → valve throat area

**orifice-plate flow meter** A differential-pressure flow meter, which consists of a thin circular plate with a concentric, eccentric or segmental hole inserted into the pipe between two flanges.

**ORP meter** ↔ oxidation-reduction-potential meter

**oscillating U-tube** A density meter, which consists of a U-shaped tube electronically excited by a piezoelectric actuator into undamped vibration. The natural frequency of such a structure is influenced only by the mass of the sample in the tube of known volume enabling the calculation of the measured density of a liquid or a gas. It is used in, e.g., chemistry, pharmacy, electronics, beverage industry, wastewater treatment. **S**: vibrating-tube density meter

**outlier** A data point, which significantly differs from other observations or measurements. It is often the consequence of various measurement errors or experimental errors. Therefore, it should be excluded from further analysis.

**output controllability** The property of a system for which such a control signal exists that transfers the system from an arbitrary initial output value to an arbitrary final output value in finite time.

**output decoupling zero** The decoupling zero of a linear MIMO system, representing an unobservable system mode.

**output disturbance** A disturbance, which is added to the controlled variable at the measuring point. It is mainly the consequence of sensor noise resulting in perturbations of the controlled signal.

**output equation** A part of the mathematical model of a lumped-parameter dynamic system in state space. It defines the relations between the state variables and the system output, as well as the direct relation between the system input and the system output. For a SISO system, it is described by an algebraic equation in the vector form. For a MIMO system, it is given in the matrix form.

**output error** The difference between the system output signal and the simulated mathematical-model output signal, where both the system and the model have the same signal at their inputs.

**output-error model** A linear model, which represents the mapping of the present output value from a finite number of previous values of inputs and outputs as well as from the present value of the input. Its coefficients are optimised according to the error of the simulated output signal. **S**: OE model

**output functional controllability** The property of a linear, functionally controllable, MIMO system that requires the zeros of the Smith-McMillan canonical form of a TFM to lie on the left-hand side of the $s$-plane. It might simplify the corresponding control design.

**output matrix** A matrix in the vector-matrix form of linear state equations of MIMO system, usually marked with letter $C$, which defines the relations between the state variables and the system outputs.

**output sensitivity function** → noise sensitivity function

**output signal** A signal that comes out of a system, e.g., the response of a dynamic system.

**output variable** A variable at the exit of the system, which changes due to an input variable or due to the dynamics of the system.

**output vector** **1**. A vector of all the variables that exit the mathematical model of a MIMO system. **2**. A vector in the vector-matrix form of linear state equations of a SISO system or of a MISO system, which defines the relations between the state variables and the system output. It is usually marked with the letter $c$. **3**. A vector, which consists of all system outputs. In the case of only one output, it degenerates into a scalar.

**over-damping** Damping, where the step response of a system reaches the new steady state without an overshoot. In such a case the damping factor is greater than 1.

**overload** The load of a system that is larger then allowed. It can result in missoperation or damage.

**oxidation-reduction-potential meter** A sensor, which measures the capacity of an aqueous system to accept or release electrons from a chemical reaction. It consists of a sensing electrode made of a noble metal and a reference electrode with a stable potential. The electrodes are submerged in the measured solution. The voltage across the circuit formed by two electrodes is proportional to the measured quantity. It is used to monitor and control chemical reactions as well as to quantify ion activity in, e.g., industry, laboratory environment. **S**: ORP meter, redox sensor

# Chapter 17
# P

**PAC** ↔ programmable automation controller

**paddlewheel level switch** → rotating paddle level switch

**Padé approximation** An approximation of dead time with a corresponding nonminimum-phase transfer function of appropriate order, which enables the simulation of a mathematical model that includes dead time.

**PAL** ↔ programmable array logic

**palletising** Robotic arranging of workpieces or products onto a tray or into a container and keeping them in an orderly sequence.

**parallel branch** → parallel path

**parallel channel** → parallel path

**parallel compensation** Compensation, in which the compensator is in the feedback path of the slave control loop. **S**: feedback compensation

**parallel manipulator** → parallel robot

**parallel path** A connection in a signal-flow graph or in a block diagram, which goes in the same direction as the forward path. It can lie along the whole forward path or along a part of it. **S**: parallel branch, parallel channel

**parallel robot** A robot, the mobile platform of which is actively linked with the robot end-effector. It is connected to a fixed base by several leg-shaped mechanisms. The structure contains active and passive joints as well as at least one closed kinematic chain. It is used in high-speed and high-accuracy positioning and pick-and-place as well as in flight simulators and automobile simulators. **S**: parallel manipulator

**parameter** A number, usually constant, which has a notable influence in functional relation, e.g., a coefficient in a parametric model.

**parameter adaptation** → self-tuning

**parameter-adaptive control** → self-tuning control

**parameter-adaptive controller** → self-tuning controller

**parameter auto-tuning** A procedure that automatically determines the parameters of the controller. The parameter values are obtained from the results of predefined and self-conducted experiments. The procedure is typically performed before the controller is put into regular operation. **S:** automatic tuning of parameters

**parameter estimation** The method for determining the values of parameters of a mathematical model from measured data, e.g., maximum-likelihood method.

**parameter optimisation** The selection of optimal parameter values for a mathematical model by maximising or minimising the objective function within preset constraints.

**parameter scheduling** → gain scheduling

**parameter sensitivity** The ratio between the change of the output variable and the change of the investigated parameter, which causes the output change.

**parametric identification method** A system identification method that results in a mathematical model with explicitly expressed parameters, e.g., least-squares method, maximum-likelihood method.

**parametric model** A mathematical model with explicitly expressed parameters, e.g., a description in state space, a description with a transfer function, a description with differential or difference equations.

**parametric uncertainty** An uncertainty, which is the consequence of the differences between parameter values of the mathematical model and their agreed true values. Such discrepancies occur mainly due to measurement problems or because of inaccuracies in the available estimates of the mathematical model parameters.

**Pareto boundary** → Pareto front

**Pareto front** A set of solutions of a multi-objective optimisation, which can not be further improved according to one criterion without being worsened according to another criterion. For two criteria used it is a curve, for three a surface, and for more a hypersurface. **S:** Pareto boundary, Pareto frontier

**Pareto frontier** → Pareto front

**Pareto optimisation** → multi-objective optimisation

**Parseval's theorem** The theorem, which describes the connection between the energy or the power of a signal defined in the time domain and the energy or the power of the same signal defined in the frequency domain. **S**: Rayleigh's energy theorem, Rayleigh's identity

**particle simulation** An approach, which enables the simulation of manufacturing processes with granular materials, liquids or suspensions expressed as groups of particles. It allows the simulation of very complex phenomena and is considered as a complement of finite-element simulation. **S**: PSO

**particle swarm optimisation** A population-based and nature-inspired stochastic optimisation method for solving continuous and discrete problems. It iteratively seeks the best solution according to a certain objective function by using a set of moving points in the search space that represent the potential individual solutions. The motion of the points depends on the previous optimal solutions and on the velocities of the individual points in the previous iteration. **S**: PSO

**partitioned method** **1**. A method, which enables the development of a simulation scheme for a mathematical model described by a differential equation that includes time-derivatives of the input variable, or by a transfer function with the numerator of order higher than 0. By introducing an auxiliary variable, the differential equation is split into two parts. The first part comprises the auxiliary variable and the output variable, whereas the second part comprises the input variable and the auxiliary variable. The combination of the two parts results in the controllable canonical form of the simulation scheme. **2**. A method, which enables the development of a simulation scheme for a mathematical model described by a difference equation that includes delayed samples of the input variable, or by a transfer function with the numerator of order higher than 0. By introducing an auxiliary variable, the difference equation is split into two parts. The first part comprises the auxiliary variable and the output variable, whereas the second part comprises the input variable and the auxiliary variable. The combination of the two parts results in the controllable canonical form of the simulation scheme.

**passive joint** A robot joint, which is not controlled and acts as a fixed robot link.

**path** **1**. A sequence of branches that conveys a signal from a source to a sink. **S**: channel **2**. The trajectory of a robot end-effector or of a mobile robot during task performance.

**path following** A robotic task, which enables a mobile robot to follow a black line on a white surface or vice-versa, as well as to follow an invisible line such as a magnetic field. A robot must be able to detect the particular line and to keep following it.

**path planning** A computational procedure in robotics, which determines a collision-free continuous path for a robot end-effector or for a mobile robot from the start position to the goal position considering diverse static obstacles. It bases on a known map of the environment. **S**: trajectory planning

**pattern-search method** → direct-search method

**payload** → load

**PCA** ↔ principal-component analysis

**P controller** ↔ proportional controller

**PD controller** ↔ proportional-differential controller

**PDF** ↔ probability density function

**peak overshoot** → maximum overshoot

**peak time** The transient-response specification defining the time, in which the transient response of a proportional underdamped second-order system to a unit-step signal reaches the extreme value for the first time.

**penalty method** An optimisation method, which replaces a constrained optimisation method by a series of unconstrained approximations whose solutions converge to the solution of the original problem, e.g., exterior penalty method, interior penalty method. The objective function is extended by adding a term that prescribes a high cost for violating the constraints.

**percentage overshoot** → maximum overshoot

**perceptron** A feedforward ANN, the output of which is a nonlinear function of weighted inputs.

**performance index** → criterion

**performance-measurement system** A system, by which organisations observe and measure their intangible performance elements in the form of qualitative or quantitative assessment. It supports decision-making in the process of constant improvement of the operation of the system under consideration.

**periodic response** A system response that repeats in equal time intervals.

**persistent excitation** The excitation of a dynamic system that enables the identification of a nonbiased mathematical model from information contained in the input signals and output signals.

**personal care robot** A service-robotics application, which performs actions contributing directly towards improvement in the quality of life of humans, excluding medical applications.

**person carrier robot** A personal care robot, which transports a human to the intended destination.

**perspective transformation** A mapping of a high dimensional space to a lower dimensional space, e.g., the projection of a 3D space into a 2D image.

**Petri net** A discrete-event model, presented mathematically or graphically, which provides the formal description of the course of actions in complex systems. It is used mainly for analyses and design of manufacturing systems, logistic systems and communication systems. For special cases, various modifications can be used, e.g., timed Petri net, coloured Petri net, stochastic Petri net. **S**: condition/event net, place/transition net

**petrol generator** A gasoline-fueled engine with spark-ignition, which powers an electric generator to provide electrical power. It is often used as a power supply of an actuator system. **S**: gasoline generator

**PFD** $\leftrightarrow$ process flow diagram

**phase** Displacement between two periodic signals with the same period, usually expressed in radians or degrees.

**phase angle 1.** $\rightarrow$ phase shift **2.** The angular component of a vector in polar coordinates, usually expressed in radians or degrees.

**phase crossover frequency** The frequency at which the phase shift is equal to $\pi$ radians, i.e., 180 degrees.

**phase difference** $\rightarrow$ phase shift

**phase distortion** The distortion of a sinusoidal signal passing through an amplifier or other dynamic system when its phase response is not linear over the frequency range of interest.

**phase lag** The phase shift for which a periodic signal is behind another periodic signal with the same period.

**phase-lag compensation** $\rightarrow$ lag compensation

**phase-lag compensator** $\rightarrow$ lag compensator

**phase lead** The phase shift for which a periodic signal is ahead of another periodic signal with the same period.

**phase-lead compensation** $\rightarrow$ lead compensation

**phase-lead compensator** $\rightarrow$ lead compensator

**phase margin** The angular difference between the stability margin, which is in $\pi$ radians or 180°, and the absolute value of the open-loop phase angle of a stable feedback system at a frequency for which the gain value is 1 or 0 dB.

**phase plane** A coordinate plane with axes representing two state variables of a dynamic system.

**phase portrait** A geometric representation of the trajectory of states of the mathematical model of a dynamic system in the phase plane, e.g., limit cycle, bifurcation, equilibrium point, chaotic attractor. **S**: phase trajectory

**phase response** The phase shift between the harmonic input-signal components and the output-response components of the system with regard to their frequencies. It is frequently represented as a part of the Bode plot of, e.g., amplifier, filter.

**phase shift** Any change that occurs in the phase of one signal, or in the difference between phase angles of two or more signals with the same period. **S**: phase angle (1), phase difference

**phase space** A multidimensional representation of a dynamic system in which each dimension corresponds to one state variable of the system.

**phase trajectory** → phase portrait

**phase variable form** → companion form

**phenomenological law** A law, which bases on experimental observations and physical insight describing some irreversible processes in the form of relation, e.g., Fourier's law of heat conduction, Fick's law of diffusion, law of chemical reactions, Ohm's resistance law.

**pH meter** A sensor, which measures the acidity or alkalinity of a solution in units 0 to 14 where at unit 7 the solution is neutral. It consists of a specially constructed probe selective to hydronium-ion concentration delivering a varying potential, and a reference electrode insensitive to hydronium-ion concentration delivering a constant potential. The electrodes, immersed in the measured solution, are connected to a high input-impedance voltmeter, which displays the results in pH units. It can be either handheld or benchtop and is commonly used in, e.g., chemistry, biology, agronomy.

**photocoupler** → optocoupler

**photodetector** A sensor, which measures the intensity of radiant energy through photoelectric action, e.g., photomultiplier tube, photodiode, avalanche photodiode, phototransistor, photovoltaic cell, bolometer. Light or other electromagnetic radiation is converted into electricity. It is used in, e.g., optical disc drive, remote control device, automatic lighting control, wireless network, fibre optics, machine vision.

**photoelectric proximity sensor** A proximity sensor, which uses the transmitted or reflected infrared or visible light to detect the presence or absence of a target object, e.g., through-beam proximity sensor, diffuse proximity sensor, retro-reflective proximity sensor. It is commonly used in industry, robotics and everyday life for, e.g., part detection on an industrial conveyor system, object detection, inspection and counting in a manufacturing system, collision detection in a robotic system. **S**: optical proximity sensor

**photometer** An optical analytical instrument, which measures various aspects of the intensity of light, e.g., illuminance, light absorption, scattering of light, reflection of light, fluorescence, luminescence. Electromagnetic radiation in the range from ultraviolet to infrared spectrum is measured through, e.g., luminous intensity, luminous flux, light distribution, colour, usually by comparing two light sources, the sample and the one with certain specific standard characteristics. It converts light into the electric signal using different photodetectors. It is used in studying the properties of liquids and solutions as well as in measuring the concentrations of organic or inorganic materials in a sample.

**physical assistant robot** A personal care robot, which helps a human to perform the required tasks by providing supplementation or augmentation of personal capabilities.

**physical model** A simplified material representation of a system, frequently on a reduced scale, which may have a static character or a dynamic character. However, its construction is often expensive, time-consuming or impractical.

**physical modelling** Computer-aided modelling of a system using graphical blocks from the corresponding libraries that deal with various problem domains. The blocks represent real components that make up the modelled system and are interconnected by lines corresponding to the connections that transmit power in the real system. Therefore, it enables a clear presentation of the system topology by describing the physical structure of the system, rather than the underlying mathematics.

**pick-and-place** The operation, in which a robot, e.g., Cartesian robot, Delta robot, collaborative robot, grabs an object in one place and drops it in another location. It is widely used in, e.g., industrial and manufacturing environments, automated storage and retrieval systems, surface mount technologies in electronics manufacturing.

**pickoff point** → branch point

**PI controller** ↔ proportional-integral controller

**P&ID** ↔ process and instrument diagram

**PID controller** ↔ proportional-integral-differential controller

**piezoelectric motor** An electric motor, which produces linear motion or rotary motion as a consequence of material deformation under the influence of the electric field. It is very small, energy-efficient, precise and can make fine steps at high frequencies. It is capable of operating in strong magnetic fields and at high temperatures. It is used in, e.g., instrumentation and control, micro-positioning, robotics, factory automation, optics, medical equipment, biotechnology.

**piezoelectric pressure sensor** An absolute dynamic-pressure sensor, which consists of quartz crystals or specially formulated ceramics that generate a charge across its faces when a force as the consequence of pressure is applied. The charge is transformed to a voltage proportional to the measured pressure. It has a rapid response to dynamic pressure changes across a wide range of pressures and frequencies.

**piezoresistive pressure sensor** An absolute pressure sensor or differential pressure sensor, which uses a semiconductor strain gauge. The strain gauge attached to a diaphragm creates the change in resistance when the diaphragm is deformed under the influence of the measured pressure.

**pilot plant** A small-scale industrial plant, which enables identification of problems as well as their solution and testing before the full-scale plant is built. It is indispensable in the refinement of new and existing industrial products as a complement to modern simulation techniques. It facilitates new technologies as well as the learning and training of personnel. In some cases, it can be used as a small production unit.

**pinch valve** A valve, which consists of a highly elastic reinforced rubber or reinforced elastomer tube in a valve body. The compression of the tube is mechanically actuated or pressure actuated with, e.g., manual actuator, solenoid actuator, electric-motor actuator, pneumatic actuator, hydraulic actuator. Its working parts are isolated from the fluid. Therefore, it can handle liquid, viscous, granular, contaminated, fibrous, or aggressive substances. It is used in, e.g., wastewater treatment, slurry handling, powder handling, pellets handling as well as in medical industry, pharmaceutical industry, food industry, plastic industry.

**pink noise** Coloured noise, the power spectral density of which is inversely proportional to the frequency. Therefore, its power spectral density decreases by 10 dB/decade. **S**: flicker noise, 1/f noise, inverse f noise

**piping and instrument diagram** $\rightarrow$ process and instrument diagram

**Pirani gauge** A very-low-pressure sensor or a vacuum sensor, which measures changes in the ability of gas to conduct heat. The rate of heat dissipation from the tungsten, nickel or platinum filament heated with constant current depends on the thermal conductivity of the surrounding medium. The consequent change in the filament temperature changes the filament resistance that is measured by a Wheatstone bridge. The obtained resistance is proportional to the measured quantity.

**pitch angle** The angle, which determines the rotation around the transversal axis of an object or a mobile system, e.g., aeroplane, submarine, robot end-effector. Besides the yaw angle and roll angle, it is an element from the set of three angles that completely determine the orientation of an object in space. **S**: elevation

**Pitot tube** A differential-pressure flow meter, which measures fluid-flow velocity only in a given point of a flowstream in pipes or in open channels. The total impact pressure is commonly measured using a small L-shaped tube with the opening oriented directly upstream, while the static pressure is measured with the probe oriented downstream or perpendicularly to the flowstream. The obtained differential pressure is proportional to the velocity of fluid that is in turn proportional to the measured flow rate.

**PLA** $\leftrightarrow$ programmable logic array

**place/transition net** → Petri net

**plan-driven process** → sequential process (2)

**plane contact** The two-dimensional contact with a surface, which has three DOF when there is no friction, and no DOF when friction exists.

**plastometer** → rheometer

**PLC** ↔ programmable logic controller

**PLD** ↔ programmable logic device

**plug valve** A valve, which uses a cylindrical or conically-tapered disc with one or more holes, rotating in the valve body to control the flow of fluid in a pipe. It is quick-acting, has a small pressure drop and leak-tight operation. It can be implemented when dealing with, e.g., air, gas, vapour, slurry, mud, sewage, as well as with high-pressure or high-temperature media.

**pneumatic capacitance** The parameter of a mathematical model, describing the property of a pneumatic assembly, e.g., storage tank, to store compressible fluid, defined as the ratio of mass flow rate to the time derivative of pressure. It is a measure of the gas capability to store potential energy. **S**: pneumatic capacity

**pneumatic capacity** → pneumatic capacitance

**pneumatic cylinder** A mechanical device, which uses the power of compressed air or gas to produce a force in reciprocating linear motion, e.g., double-acting cylinder, telescopic cylinder, rodless cylinder. It is used in, e.g., pulse tool, hand-held tool, elevator, garbage truck. **S**: air cylinder

**pneumatic inertance** 1. The parameter of a mathematical model, describing the property of moving compressible fluid in a pneumatic assembly, e.g., pipe, defined as the ratio of pressure that accelerates or decelerates the gas to the time derivative of volumetric flow rate. It is the measure of inertia of the moving gas, which is mostly negligible. 2. The parameter of a mathematical model, describing the property of moving compressible fluid in a pneumatic assembly, e.g., pipe, defined as the ratio of pressure that accelerates or decelerates the gas to the time derivative of mass flow rate. It is the measure of inertia of the moving gas, which is mostly negligible.

**pneumatic motor** A mechanical device, which converts compressed-air power or gas power to either linear or rotary motion, e.g., piston motor, vane motor, gear motor, turbine motor. It is quiet, clean and sparkless and therefore frequently used in an explosive, hazardous environment or where extreme cleanliness is required. **S**: air motor

**pneumatic proximity sensor** A proximity sensor, which uses a compressed air jet to detect the presence of an object without touching it. The measured pressure changes due to the reflection of the air jet from the surface of the detected object. It is often used in food processing and beverage processing. **S**: back-pressure proximity sensor

**pneumatic relay** A gas-flow amplifier, which consists of two chambers, one connected to the gas supply and the other connected to the output. They are separated by a flexible membrane. The latter affects the valve that controls the output flow. It is an important element in the automatic control of pneumatic systems. **S**: air relay

**pneumatic resistance** **1**. The parameter of a mathematical model, describing the property of a pneumatic assembly to resist compressible-fluid laminar flow, defined as the ratio of pressure drop to volumetric flow rate. **2**. The parameter of a mathematical model, describing the property of a pneumatic assembly to resist compressible-fluid laminar flow, defined as the ratio of pressure drop to mass flow rate.

**point contact** The zero-dimensional contact with a surface, which has five DOF when there is no friction, and three DOF when friction exists.

**point-to-point control** Control of the robot, which moves the robot end-effector from the start position to the next position. The intermediate path is determined by the robot controller. **S**: PTP control

**polar diagram** $\rightarrow$ polar plot

**polar plot** A diagram depicting the frequency response of a system in the complex plane. Each point of the diagram represents the frequency response at a particular frequency. The frequency is thus the parameter, ranging from 0 to infinity. It is used in control-system analysis and design. **S**: polar diagram

**polar robot** A robot, which has two rotational joints and one prismatic joint. Therefore, the outer limit of its reachable workspace is defined by the radius from the intersection of the two rotational-joint axes to the robot end-effector with a fully extended prismatic joint. It is often used in, e.g., die casting, handling of machine tools, gas welding, arc welding. **S**: spherical robot (2)

**pole** **1**. A value of complex variable $s$ or complex variable $z$, which results in a zero-valued denominator of the continuous or discrete transfer function and therefore makes the transfer function singular. Its position in the complex plane affects the stability and behaviour of the system. **2**. An eigenvalue of the system matrix. Its position in the complex plane affects the stability and behaviour of the system. **3**. A root of the characteristic equation of a linear closed-loop system. Its position in the complex plane affects the stability and behaviour of the system.

**pole allocation** **1**. $\rightarrow$ full-state feedback **2**. $\rightarrow$ pole assignment (2)

**pole assignment** **1**. $\rightarrow$ full-state feedback **2**. A family of control-design methods for linear MIMO systems. The parameters of the controller are calculated according to the requirements concerning the predefined locations of some or all closed-loop system poles in the $s$-plane. A controller that uses state feedback or output feedback can be designed. However, the choice of closed-loop poles that ensures the desired control-system behaviour might be complex. **S**: eigenvalue assignment (2), pole allocation (2), pole placement (2), pole shifting (2)

**pole excess** → relative degree

**pole placement** **1**. → full-state feedback **2**. → pole assignment (2)

**pole-placement controller** → full-state feedback controller

**pole shifting** **1**. → full-state feedback **2**. → pole assignment (2)

**polynomial transfer-function form** The structure of a transfer function, which has polynomials of complex variable $s$ or complex variable $z$ in its numerator and denominator.

**population dynamics** A field of life sciences, dealing with biological, ecological and sociological phenomena, e.g., birth rate, death rate, immigration, emigration, which are described by a population model.

**population model** A mathematical model, which describes growth, stability and decline of a community of humans, animals or plants, e.g., Malthusian growth model, Gompertz demographic model, Lotka-Volterra predator-prey model, Bertalanffy model, matrix population model.

**pose** Position and orientation of a body, described in space by six DOF relative to the predefined coordinate frame.

**position** **1**. The location of an object according to the origin of the coordinate system. **2**. The distance of an object from the origin of the coordinate system along a particular path. **3**. Three translational DOF describing the location of an object in space.

**position control** Control of the robot joint variables, in which the reference signal is the desired pose of the robot end-effector.

**positioner** A device, which assures the desired displacement of the moving part of an actuator, often valve actuator, according to the electric or pneumatic control signal. It is the interface between the control system and the actuator ensuring the reliability of the controlled process.

**position-error constant** **1**. A closed-loop system parameter, which is defined by the limit value of the open-loop transfer function as $s$ approaches 0. Here, $s$ is the complex variable in the $s$-plane. **2**. A closed-loop system parameter, the value of which increased by one is inversely proportional to the steady-state error of the response of the system to a unit-step signal.

**positive-displacement flow meter** A flow meter, which measures the volumetric flow rate of gas, liquid or viscous liquid. It consists of a chamber that obstructs the flow, and of a rotating or reciprocating mechanism that allows a fixed-volume amount of media to pass from the input to the output of the device, e.g., rotary-vane flow meter, gear flow meter, reciprocating-disc flow meter. The number of displacements is proportional to the measured flow rate. It is used in, e.g., oil plant, petroleum plant, gas plant, water treatment, wastewater treatment, food production, beverage production, chemical industry, pharmaceutical industry, cosmetic industry.

**positive displacement pump** A pump, which moves fluid by trapping a fixed amount of fluid and forcing the trapped volume into the discharge pipe, e.g., gear pump, rotary vane pump, radial piston pump, axial piston pump, diaphragm pump. It provides an approximately constant flow at a fixed speed despite the changes in counter pressure.

**positive feedback** A feedback, in which the setpoint value is added to the output value. It is in phase with the input, tending to push the output away from the desired value. It has a destabilising effect that results in exponential growth, oscillations or chaotic behaviour of the system and amplifies the impact of disturbances and noise. It appears in, e.g., audio technics, electronic engineering, chemistry, biology, cybernetics, economy, climatology, sociology, psychology. **S**: regenerative feedback, reinforcing feedback

**potentiometer** A displacement sensor, which measures the linear motion of an object connected to a sliding element of the voltage divider converting it into an electrical signal. The change in position causes the resistance change between the fixed point and slider, which is proportional to the measured displacement.

**power amplifier** An amplifier, which provides energy to the systems connected to its output, e.g., transistor power amplifier, magnetic amplifier, spool valve, lever, gears. The energy is provided by the corresponding power supply.

**power converter** An electrical device which performs the necessary conversion of the control signals, e.g., pulse-width modulator, variable-speed drive, inverter.

**power cycle** $\rightarrow$ duty cycle (1)

**power density spectrum** $\rightarrow$ power spectral density

**powered exoskeleton** $\rightarrow$ robotic exoskeleton (1, 2)

**power grid** An interconnected network, which delivers electrical energy from producers to consumers. It is often used as a power supply of an actuator system. **S**: electrical grid

**power source  1**. A source, from which useful energy can be extracted or recovered, either directly or by conversion or transformation process, e.g., solid fuels, liquid fuels, solar energy, biomass. **S**: energy source **2**. A device or system, which stores, supplies or converts energy that enables a power amplifier to provide the power needed by an actuator or by a final control element, e.g., rechargeable battery, fuel cell, diesel generator, power grid, compressor, pump. **S**: energy source

**power spectral density** A function describing a frequency-dependent representation of signal power. The integral of the function along an arbitrary frequency interval is a constituent part of the whole power of the signal. It is suitable for describing signals with finite power, e.g., noise. **S**: power density spectrum, power spectrum

**power spectrum** $\rightarrow$ power spectral density

**power supply** An electrical device, which receives energy from a corresponding power source converting one type of electrical power to another, e. g., DC power supply, AC power supply, programmable power supply, high-voltage power supply, UPS. It also converts different forms of energy into electrical energy. It is used in, e.g., automation, computers, electric vehicles, aircraft, welding, medical equipment.

**pre-act time** → derivative time

**preamplifier** An electrical circuit, which processes the control signal to make it compatible with the power-amplifier input. It is usually implemented using an operational amplifier. It is often an element of an actuator system.

**precision** **1**. A description of random errors representing a measure of statistical variability. **2**. Proximity of the repeated measurements of the same variable to each other, expressing fineness to which a measuring instrument can be repeatedly and reliably read. **3**. The property of a model where the output it provides is a definite value rather than a range of values, or a definite mathematical function rather than a family of mathematical functions.

**prediction estimator** → Luenberger observer

**prediction horizon** A parameter in MPC denoting the number of future time steps that are considered when seeking the optimal control variable in a particular time step. It can be greater than or equal to the control horizon.

**prediction observer** → Luenberger observer

**predictive validity** A model-validation procedure, which determines the degree to which the model response fits the test data. It rises the confidence in the ability of the model response to predict the behaviour of the modelled system.

**predictor-corrector integration method** A numerical integration method, which uses different combinations of single-step methods and multi-step methods to improve its properties.

**premise** → antecedent

**pressure gauge** → pressure sensor

**pressure-reducing valve** A valve, which sets the input pressure of gas or liquid in a system to the desired output pressure, regardless of a changeable flow rate or variable input pressure. **S**: pressure-reduction valve, pressure-regulating valve

**pressure-reduction valve** → pressure-reducing valve

**pressure-regulating valve** → pressure-reducing valve

**pressure sensor** A sensor, which measures either absolute pressure or differential pressure of liquids and gases, e.g., U-tube manometer, diaphragm pressure sensor, bellows pressure sensor, Bourdon tube, piezoelectric pressure sensor, MEMS pressure sensor, optical pressure sensor. It is one of the most common sensors. It often serves as a transducer in measurement systems for, e.g., flow rate, temperature, level, velocity, altitude. **S**: pressure gauge

**pressure spring thermometer** A temperature sensor, which utilises volumetric expansions of a thermally-sensitive fluid caused by temperature changes, i.e., liquid-filled thermometer, gas-filled thermometer. It consists of a metal bulb containing fluid inserted at the spot of measurement and a metal capillary that transmits the pressure in the bulb to a receiving element. The latter converts pressure into displacement, indicating the measured temperature on a calibrated scale. **S**: Bourdon tube thermometer, filled system thermometer

**pressure switch** A device, which changes its binary output at the moment its pressure preset value is reached, e.g., mechanic pressure switch, electronic pressure switch. It consists of a pressure sensor and a switching contact. It is commonly used in general industrial applications, as well as in, e.g., engine monitoring, machine tools, medical equipment.

**primary battery** A portable voltaic cell, which is not rechargeable, e.g., alkaline battery. It is often used as a power supply of an actuator system.

**primary controller** → master controller

**primary control loop** → master control loop

**principal-component analysis** A statistical method, which determines the most significant variables, containing the essential information about the process behaviour. The observed variables are transformed into linearly independent principal components that create uncorrelated orthogonal basis set. **S**: PCA

**principal gain** → singular value (2)

**principal value** → singular value (2)

**principle of parsimony** → Occam's razor

**prismatic joint** A single-DOF robot joint, which constrains the movement of two neighbouring robot links to displacement along a line. The relative position of one robot link with respect to the other robot link is given by the distance along the robot-joint axis. **S**: translational joint

**probabilistic model 1**. → stochastic model **2**. A stochastic model, where changes in its output responses in each time instant are described by the same probability distribution functions as in the past, with no reminiscence of responses' past.

**probabilistic system** → stochastic system

**probability density function** A function that describes the probability distribution of a continuous random variable, e.g., noise source. The probability of the random variable falling within an interval between two particular values is equal to its definite integral over that interval. **S**: PDF

**probability distribution function** A function that describes the probability of occurrence of different possible values of a random variable.

**problem domain 1**. The area of expertise or application, which needs to be examined to solve a posed problem. **2**. The professional field, in which the problem or application, e.g., modelled system, originates.

**problem-oriented simulation language** → special-purpose simulation language

**procedural control** The higher level of process-control level of the CIM, which implements technology-aimed actions for larger technological operations with commands to the basic control and with monitoring of the response from the basic control. It is usually implemented as supervisory discrete-event control or as sequential control of, e.g., substance-dosage phase in the food industry.

**process** A set of mutually dependent actions in a system that result in transformation, transport or storing of matter, energy or information.

**process and instrument diagram** A detailed graphical representation used in the process industry, which shows process equipment together with instrumentation and control devices using standard symbols. **S**: P&ID, piping and instrument diagram

**process-behaviour chart** → control chart

**process control** A control engineering discipline that deals with structures, mechanisms and algorithms, which enable the implementation and maintenance of the prescribed behaviour of the controlled system. In industrial control systems, it assures production-level consistency, economy and safety, which could not be achieved by a human, or using manual control. It is used in, e.g., chemical industry, power generation industry, oil refining industry.

**process controller** → industrial controller

**process-control level** The low level of the CIM, which comprises control of technological processes and procedures in an industrial company. It consists of basic control and procedural control.

**process fieldbus** An open, supplier-independent, master-slave serial fieldbus, the interface of which permits various applications in process automation and production automation, e.g., PROFIBUS PA, PROFIBUS DP. **S**: PROFIBUS

**process field net** An industrial Ethernet for connecting equipment in industrial automation applications, which can deliver data under tight time constraints for exchanging data, alarms and diagnostics among various controllers and for configuring, accessing and controlling devices in a distributed automation system, e.g., PROFINET IO, PROFINET CBA. **S**: PROFINET

**process flow diagram** A graphical representation, which shows the relationship between major components in an industrial plant using the corresponding symbols and arrows showing the direction of flow. It tabulates the design values in different operating modes, e.g., minimum values, normal values, maximum values, not showing any minor components, e.g., bypass line, maintenance vent, shut-off valve. It is used in process industry and chemical industry. **S**: PFD

**process parameter** → process variable

**process value** → process variable

**process valve** A valve intended for the flow-control of fluids, e.g., ball valve, butterfly valve, control valve, on-off valve, flap valve. It is used in numerous industrial and nonindustrial applications.

**process variable** A current measured value in the monitored or controlled system, which shows its status, e.g., temperature, pressure, flow, level. It indicates whether it meets the plan or some adjustments are needed. **S**: process parameter, process value

**production-control level** The middle level of the CIM, which comprises activities of monitoring, controlling, planning, maintenance and scheduling of production, transport, and QC. **S**: manufacturing operations and control

**production informatics** Usage of computational technologies in industrial systems, connecting the control level with the management level of the automation pyramid.

**production management** The process of planning, organising, directing and supervising production activities from raw material to finished products and services. It brings together human resources, financial resources, machines, materials, methods and market. It enables decision-making regarding scheduling, cost, performance, quality and waste requirements according to specifications. It helps to generate employment, improve quality, face competition, introduce new products, satisfy customers, expand the firm and increase its reputation and image.

**production scheduling** A set of activities including planning, routing, organising time and dispatching, which, according to the plan, define the optimal use of resources to fulfil the requirements on time.

**production system** 1. A process that transforms resources into goods and services. 2. A system in which individual workpieces are successively processed in appropriate intervals in, e.g., wood processing, metal processing.

**product life cycle** → system-development life cycle

**PROFIBUS** ↔ process fieldbus

**PROFINET** ↔ process field net

**profit function** → fitness function

**prognostics** An engineering discipline that uses fault-diagnosis data from the past to enable the prediction of the time, at which a system or component will no longer perform its specified function. It supports the decision about the needed maintenance or contingency mitigation to assure the normal operation of a system.

**program** A set of instructions, scripts and data for executing a specific task.

**programmable array logic** PLD, the output of which is a non-programmable OR-gate array that is connected to an input-driven programmable AND-gate array. It can be further connected to other similar circuits. It is easily manufacturable and exhibits a good speed performance. **S**: PAL

**programmable automation controller** An open architecture, modular, high performance, feature-rich industrial controller, which incorporates high-level instructions, merging the features of a PLC, a data logger, a communication gateway, and an embedded computer. It is used to communicate, monitor and control processes across multiple networks and devices, utilising standard protocols and network technologies. **S**: PAC, softPLC

**programmable logic array** PLD, the output of which is a programmable OR-gate array that is connected to an input-driven programmable AND-gate array. It can be further connected to other similar circuits. It is a flexible, compact and space-efficient solution for complex circuits often used in control systems. **S**: PLA

**programmable logic controller** A controller, which is specially designed to operate reliably in harsh environments, which is crucial in the field of industrial automation. It can process information, has no mechanical parts, takes up a small space, needs low power, performs complex tasks, and is extremely customisable. It can be programmed according to the operational requirement of the process, usually using an appropriate standard PLC programming language. It can easily be reprogrammed when the operational requirements change. **S**: PLC

**programmable logic device** A semiconductor-based electronic component, which enables the building of reconfigurable digital circuits, e.g., PAL, PLA, CPLD, FPGA. Before its implementation, it must be programmed using specialised software and is, therefore, the combination of processing and memory parts, e.g., EPROM memory cell, EEPROM memory cell, flash memory. It is flexible, customisable, modifiable, small-sized, has low power consumption and is thus used in various industrial applications. **S**: PLD

**programmed control** Control of a system according to a predetermined program. It can be achieved with feedback control or open-loop control, e.g., control of in-advance-scheduled flight trajectory.

**programmer** **1**. A device, which automatically controls the operation of a certain apparatus according to the prescribed program. **2**. Electronic equipment that arranges written software to be loaded into devices, e.g., microcontroller, PLD, PAL, PLA, FPGA. **3**. A person who creates specifications for computer software, which is usually implemented by a coder.

**programming by demonstration** → imitation learning

**proper Euler angles** Euler angles, which determine the orientation of an object in space with regard to three sequential rotations around two different axes of the coordinate system. The sequence of rotation around axes $x$, $y$ and $z$ can be one of the following: $z$-$x$-$z$, $x$-$y$-$x$, $y$-$z$-$y$, $z$-$y$-$z$, $x$-$z$-$x$ or $y$-$x$-$y$. **S**: classic Euler angles

**proper model** A model, which is not more complex than necessary for the given purpose.

**proper transfer function** A transfer function, in which the degree of the numerator does not exceed the degree of the denominator, i.e., strictly proper transfer function or biproper transfer function. Its relative degree is greater than 0 or equal to 0. It describes a causal system, the response of which never grows unbounded as the frequency approaches infinity.

**proportional band** A percentage of error change, which causes the final-control-element output to change from the lowest to the highest value. It is inversely proportional to the proportional gain.

**proportional control** A control strategy, in which the controller output is linearly dependent on the error, which is the controller input. **S**: proportional-control action

**proportional-control action** → proportional control

**proportional controller** A controller, the output of which is the error multiplied by a constant. Its input is the error, whereas its output is the control signal. It reduces steady-state error and speeds up closed-loop system response. **S**: P controller

**proportional-differential controller** A controller, the output of which is a weighted sum of the error and the derivative of the error. Its input is the error, whereas its output is the control signal. **S**: PD controller

**proportional gain** A constant, which is the factor between controller output and error. It is used in feedback control with the controller that contains a P term.

**proportional-integral controller** A controller, the output of which is a weighted sum of the error and the integral of the error. Its input is the error, whereas its output is the control signal. **S**: PI controller

**proportional-integral-differential controller** A controller, the output of which is a weighted sum of the error, the integral of the error, and the derivative of the error, with possible industrial modifications and realisations. Generally, its input is the error, whereas its output is the control signal. **S**: PID controller, three-mode controller, three-term controller

**proportional system** A system, the dynamics of which can be described by a transfer function with no poles or zeros in the origin of the $s$-plane. Its step response settles at a constant steady-state value, which is different from the initial value of the response.

**proportional term** A subsystem, the output of which is obtained by multiplying its input signal by a constant. It is usually a constituent part of a controller. **S**: P term

**proprioception 1**. Ability to sense self-movement and body position accurately knowing the positions and movements of skeletal joints. **S**: kinesthetic sense **2**. Ability to sense stimuli arising in the body regarding position, motion and equilibrium. **S**: kinesthetic sense

**protective separation distance** The shortest permissible distance between any moving hazardous part of the robot system and a human in a collaborative workspace.

**protective stop** The interruption of operation that terminates motion for safeguarding purposes enabling facilitated restart.

**prototype 1**. A physical model with a dynamic character, usually representing a reduced-scale copy of the whole process, of its parts, or of the device, e.g., laboratory setup, pilot plant. **S**: mockup **2**. A product, which is intended for testing or upgrading before its serial production. **S**: mockup

**proximal direction** A direction away from the robot end-effector towards the robot base.

**proximity sensor** A contactless discrete sensor, which detects the presence of an object in a certain region around it, e.g., capacitive proximity sensor, inductive proximity sensor, photoelectric proximity sensor, pneumatic proximity sensor, ultrasonic proximity sensor, Hall-effect proximity sensor. It is commonly used in, e.g., manufacturing, food production, recycling, robotics, fluid detection, conveyor systems, parking systems.

**pseudorandom-number generator** → random-number generator

**PSO** ↔ particle swarm optimisation

**psychrometer** A humidity sensor, which consists of a temperature sensor with a dry bulb and a temperature sensor, the bulb of which is kept wet by a thin wet cloth wick. The evaporation caused by the flow of the measured gas, to which the bulbs of both temperature sensors are exposed, generates a temperature difference due to evaporation and consequently a lower temperature of the wet temperature sensor. The corresponding tables enable the determination of relative humidity and dew-point temperature. It is used in, e.g., heating, ventilation, air-conditioning, meteorology.

**P term** ↔ proportional term

**PTP control** ↔ point-to-point control

**pulse train** A signal, which consists of several pulses that are mutually separated by fixed and often constant time intervals. The duration of each pulse and its amplitude are also often equal.

**pulse-width modulation** A technique that generates variable-width fixed-amplitude pulses to represent the amplitude of the analogue modulated signal. It enables control of power, which is delivered to the load, e.g., electric motor, without dissipating any wasted power. **S**: PWM

**pump** A device, which uses mechanical action to raise, transport, deliver or compress fluids or slurries, e.g., positive displacement pump, gear pump, centrifugal pump. It adds energy to a fluid to increase its flow rate and static pressure. This is enabled by different power sources, e.g., electrical motor, the steam engine, wind power, manual operation. It is often used as a power supply of an actuator system and can be used in various pneumatic or hydraulic systems and applications.

**pure number** → dimensionless quantity

**purple noise** → violet noise

**PWM** ↔ pulse-width modulation

**pycnometer** A density meter, which weighs a given volume of a low-viscosity liquid, a medium-viscosity liquid or a granular solid material. The measured density can be calculated from the difference between the masses of an empty and filled bottle with a known volume or by comparing the weight of the medium with known density and the weight of the medium under test, both in the same bottle of unknown volume. It is used as a laboratory device for, e.g., QC, research, development, testing, calibration or in, e.g., cosmetic industry, paint industry, food industry.

**pyrometer** A noncontact temperature sensor for distant objects, which detects the emitted electromagnetic radiation, i.e., optical pyrometer, infrared pyrometer. A light-sensitive photodetector or thermal detector converts the measured radiation to the corresponding electrical signal. It can measure a wide range of temperatures of objects from a greater distance or moving objects sometimes using also optical fibres. It is used in, e.g., metallurgy, smelting furnace, steam boiler. **S**: radiation thermometer

# Chapter 18
# Q

**QA** ↔ quality assurance

**QC** ↔ quality control

**quadcopter** A drone with two propellers rotating clockwise and two propellers rotating counterclockwise. It is equipped with sensors, e.g., gyroscope, accelerometer, digital camera, collision-avoidance sensor, and controlled by changing the angular velocity of the corresponding propellers. It is usable in, e.g., surveillance, research, delivery, military tasks, humanitarian operations. **S**: quadrotor

**quadrotor** → quadcopter

**qualitative model** A simple mathematical model, obtained from a description of the model structure and its parameters without stating their numerical values. It is often used when limited data are available or when only a rough estimation of the system behaviour is required. It can also be used in artificial intelligence model-design methods.

**quality assurance** A procedure, which aims to prevent the causes of inadequate quality of products and services. It is based on the results of fault diagnosis and is performed before QC. **S**: QA

**quality control** Ongoing supervision of a production process, which is focused on the maintenance of the specified measurable characteristics of the product or service that are determined according to the requirements of the users as well as to the available standards. It tends to prevent the defects at the source using efficient feedback and corrective actions procedure. **S**: QC

**quantisation** **1**. A process of mapping values from a larger set to a smaller set with a finite number of elements. **2**. The process of approximating a continuous signal by a finite number of values. It is often used in digital signal processing.

© ZRC SAZU/Research Centre of the Slovenian Academy of Sciences and Arts 2023  171
R. Karba et al., *Terminological Dictionary of Automatic Control, Systems and Robotics*,
Intelligent Systems, Control and Automation: Science and Engineering 104,
https://doi.org/10.1007/978-3-031-35755-8_18

**quantisation error** The difference between the quantised value, often of a digital signal, and the actual value of the analogue signal. It is caused either by rounding or truncation in A/D conversion and depends on the resolution of the A/D converter.

**quantisation noise** The noise that is a consequence of the quantisation error that occurs when converting an analogue signal to a digital signal. Its power decreases by increasing the resolution of the A/D converter.

**quantity of dimension 1** → dimensionless quantity

**quartz thermometer** A temperature sensor with a high resolution, which infers the temperature from the resonant frequency of a specially-cut crystal. The corresponding crystal oscillator exhibits a linear relationship between its resonant frequency and temperature. It is accurate, long-term stable and reliable in a relatively narrow temperature range. It is frequently used for calibrating other temperature sensors.

**quasi-static contact** A contact, in which the operator body part is clamped between a part of a robot system and another fixed or moving part of a robot cell.

**quaternion** A generalisation of complex numbers, which is represented by four real numbers. It describes the orientation of an object in space, e.g., in robotics.

**queueing system** A data structure, which mimics the real-world queues that occur when customers demand services from some facility in, e.g., bank, supermarket, parking lot, production system, hospital's emergency department. The arrivals of customers and service times are assumed to be random.

**quick-opening characteristic** An inherent valve characteristic, which ensures the maximum change in flow rate for small valve travels. Further increase in valve travel causes a nearly negligible change in the flow rate. It is frequently used for a valve in the on-off regime or in, e.g., safety system, cooling water system, where a large flow rate may be instantly needed. **S**: fast opening characteristic

# Chapter 19
# R

**RAD** ↔ rapid application development

**radar level sensor** A level sensor, which bases on the time-of-flight of electromagnetic waves. The material surface reflects the high-frequency electromagnetic waves emitted by an antenna. A detector receives the reflected waves enabling the calculation of the measured level from the time-of-flight and known geometry of the container. Nonconductive target material must have a sufficient dielectric constant to assure adequate reflection. It enables noncontact continuous measurements in various applications. **S**: radar level transmitter

**radar level transmitter** → radar level sensor

**radial basis-function network** An ANN that uses real-valued functions whose values depend only on the distance between some fixed point and function arguments. These functions are used as the functions for nonlinear mapping between inputs and outputs.

**radial centrifugal pump** A pump, which consists of a rotating impeller in a pipe that causes the increase of the pressure and flow of a fluid. The latter is pumped perpendicularly to the impeller shaft and strongly exploits centrifugal force. It is used in, e.g., process industry, water treatment, air conditioning, plastics industry, cooling, refrigeration.

**radial piston pump** A positive displacement pump, which consists of several pistons moved by an eccentric shaft transferring liquid or gas relatively smoothly from inlet port to outlet port. It is capable to produce extreme pressure. It is used in, e.g., plastic and powder injection moulding, machine tools, wind energy, automotive industry.

© ZRC SAZU/Research Centre of the Slovenian Academy of Sciences and Arts 2023
R. Karba et al., *Terminological Dictionary of Automatic Control, Systems and Robotics*,
Intelligent Systems, Control and Automation: Science and Engineering 104,
https://doi.org/10.1007/978-3-031-35755-8_19

**radiation detector** A sensor, which measures, tracks, or identifies ionising particles that are produced by cosmic radiation, nuclear decay, collisions in a particle accelerator, e.g., ionisation chamber, proportional counter, Geiger-Müller tube, scintillation detector, solid-state detector. It can be used in, e.g., radiation monitoring, radiation measuring, radiation protection, radiation search, radiation dosimetry, radiation security.

**radiation level sensor** A level sensor, which bases on the attenuation of gamma rays as they penetrate liquid or solid materials. It is noninvasive and noncontacting. The gamma source and the corresponding detector can be mounted in different positions inside or outside of the container enabling adaptation to the needs of measurement. It is used in a wide range of temperatures, pressures, viscosities, chemical characteristics of the measured fluid and is unaffected by turbulence, mist, foam, heavy vapour. **S**: nuclear level sensor

**radiation thermometer** → pyrometer

**radioactive ionisation gauge** → alphatron vacuum gauge

**ramp function** The function for modelling a test signal, which is 0 for negative values of the independent variable and linearly proportional to positive values of the independent variable.

**ramp signal** An aperiodic standard signal for testing the behaviour of dynamic systems, which is modelled by a ramp function.

**random-number generator** Any algorithm, program or simulation language module, which produces seemingly coincidental sequences of symbols or numbers, usable in, e.g., discrete-event simulation, Monte Carlo simulation, security, computer games, gambling. As digital computers are deterministic and entirely predictable devices, they cannot generate truly coincidental sequences. The sequences are repeatable and completely determined by the seed, i.e., the number used to initialise the algorithm. **S**: pseudorandom-number generator

**rangefinder** → distance sensor

**range sensor** A sensor, which captures the 3D structure of the surrounding. The measurements of the distance to the nearest surface can be at a single point, across a scanning plane, or a full image with measurements at every point. It allows the robot, e.g., to find navigable routes more reliably, to avoid obstacles, to grasp objects.

**rapid application development** An iterative approach to building software applications, which emphasises adaptability and speed of development. It reduces the risk as well as the development and maintenance costs, increases the quality of the application and enables faster time to market. However, highly skilled developers and strong team collaboration must be assured. **S**: RAD

**rapid control prototyping** A methodology, which accelerates the control-development process as well as the control-strategy testing on a physical system in the early phase of control design. It often offers the on-line validation of design results. The corresponding tools provide significant support to engineers in solving real-life control problems. **S**: RCP

**rapid prototyping** A group of technologies for making a physical model of the system or its part, usually using CAD data and 3D printing.

**rate** → flow

**rate of fluid flow** → volumetric flow rate

**ratio control** A control structure, which enables the maintenance of the relationship between two variables to control a third variable, e.g., mixing of two fluids. Only one of the two variables is influenced by a control signal.

**rational canonical form** → companion form

**Rayleigh's energy theorem** → Parseval's theorem

**Rayleigh's identity** → Parseval's theorem

**RCP** ↔ rapid control prototyping

**reachability** The property of a system, for which a control signal exists such that any arbitrary state can be reached in finite time from any initial state. For a continuous-time linear system, it is equivalent to controllability. A corresponding discrete-time system always exhibits controllability.

**reachable workspace** The area, which can be reached by the robot end-effector. **S**: accessible workspace

**reaching time** The transient-response specification defining the time, in which transient response of a proportional underdamped second-order system to a unit-step signal enters a prespecified range around the steady-state value for the first time. The range is given as the prespecified percentage of the steady-state value. Frequently used values are 5 % or 2 %.

**reaction curve** The s-shaped step response of a linear damped or overdamped proportional higher-order system. See Fig. 3.

**reaction-curve method** → step-response method

**reaction rate** The parameter used for tuning PID-type controllers, e.g., with the step-response method, which is defined as the slope of the tangent at the inflection point of the reaction curve. See Fig. 4.

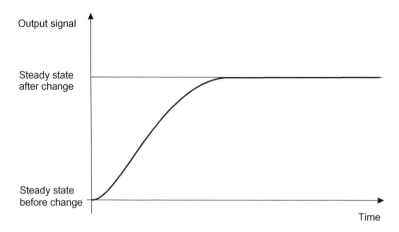

**Fig. 3** Reaction curve

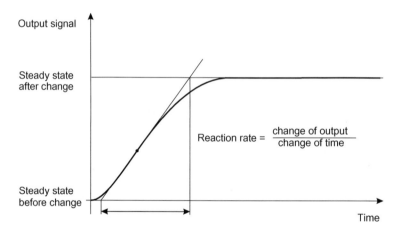

**Fig. 4** Reaction rate

**reaction time** 1. A dynamic property of a measurement system, which is described by its time constant. 2. A dynamic property of a measurement system, which is described by its bandwidth.

**reaction torque sensor** A torque sensor, which converts static torsional mechanical input into an electrical output signal. It most often uses metal foil strain gauges bonded onto a deformed mechanical component.

**reaction wheel** A flywheel, which is used to influence spacecraft orientation by changing its rotational speed and thus generating torque. Therefore, it replaces thrusters or external applications of torque. **S**: RW

**reactive agent** An agent that perceives its environment and responds to it in a timely fashion according to a preset action defined by a designer, by learning, by evolution or by their combination. It can express complex behaviour by using simple components and rules, but it is unable to remember, plan or reason logically.

**realisability** A property of a project, for the realisation of which all the neccessary conditions are fulfilled, e.g., financial constraints, ecological requirements, safety regulations, level of technological development.

**real-time control system** A digital control system with internal time-rate that is synchronised with the time-rate as sensed by humans.

**real-time operation** An operation that is carried out at the same time-rate as sensed by humans, e.g., real-time simulation, real-time control system, real-time computing.

**real-time simulation** A simulation, in which time, as an independent variable of the simulation model, is synchronised with the time of the modelled system. Consequently, it runs at the same rate as the physical system.

**real-time system** System that operates in a time as sensed by humans.

**receding-horizon control** $\rightarrow$ model predictive control

**rechargeable battery** A portable voltaic cell, which can be charged multiple times. It uses several combinations of electrode materials and electrolytes, e.g., lead-acid, nickel-cadmium, lithium-ion, lithium-ion polymer. It is often used as a power supply of an actuator system. **S**: accumulator battery, secondary battery

**recipe control** Control, which enables implementation and simple changes in the sequence of steps needed in the manufacturing of certain chemical or biological substances. It is usable particularly in plants, where a wide range of products is produced using the same or similar process sequences, as well as in batch processes.

**rectangular pulse** A unidirectional nonsinusoidal test signal, which is modelled with a rectangular pulse function. It is one of the most commonly used test signals because it is a basic element of all digital signals as well as of signals in control theory.

**rectangular pulse function** A function, which models the test signal with a rectangular shape of specified duration and amplitude.

**rectangular robot** $\rightarrow$ Cartesian robot

**rectifier** An electrical circuit, which converts AC power to DC power. It is used in various devices, e.g., DC power supply, smartphone charger, notebook. **S**: AC/DC converter

**rectilinear motion** $\rightarrow$ linear motion

**recurrent neural network** An ANN where some or all of its elements are interconnected and also contains feedback loops, e.g., Hopfield neural network, Boltzmann machine. It is frequently used for describing the behaviour of time series.

**recursive identification method** The identification method where the mathematical model or its parameters in the next time step are computed from the model or its parameters in the current time step, and from new observations in each time step.

**red noise** → Brownian noise

**redox sensor** → oxidation-reduction-potential meter

**reduced-order observer** → minimum-order observer

**redundant manipulator** **1**. A manipulator, which is made up of more robot joints than the minimum required to execute its task. It has an increased level of dexterity. **S**: redundant robot **2**. A manipulator, which has more robot joints than the minimum required for the robot end-effector to reach a certain pose. It allows its configuration to change while its robot end-effector remains in the fixed pose. **S**: redundant robot

**redundant robot** → robot manipulator (1, 2)

**reed relay** A proximity sensor, which detects the presence of a magnet or an object with attached magnet. It consists of magnetic contacts sealed in a narrow glass tube. The contacts react when a magnet comes in a certain proximity of the sensor. It can also be used as a switch.

**reference** → setpoint

**reference input** → reference signal

**reference signal** The signal, which sets the prescribed value for the control-system output. **S**: reference input

**reference tracking** → setpoint tracking

**reflux valve** → check valve

**refractometer** An analytical instrument, which measures the refractive index of gases, liquids or translucent solids, e.g., digital handheld refractometer, Abbe laboratory refractometer, inline process refractometer. The light passes through different media at different velocities, causing the light beam to change direction at the interface between two adjoining materials. The critical angle, at which the light beam is neither bent nor reflected, is measured and correlated with the refraction index and in turn with the concentration of the solution. Automatic temperature compensation is often included. Measurements in, e.g., suspension, emulsion, mixture, slurry, enable the determination of the content of substances in, e.g., wine, juice, jelly, honey, paste, electrolyte solution, paint.

**regenerative feedback** → positive feedback

**regression model** A mathematical model, which uses a statistical method to obtain the functional relation between the output variable and the input variables of a system. It is used for the prediction of the output variable of a system.

**regulator** → controller (1)

**rehabilitation robotics** A subfield of service robotics, in which robots assist disabled persons to complete the necessary activities, or provide therapy for persons, aiming to improve their physical or cognitive functions, e.g., prosthetic robots, therapy robots.

**reinforcement learning** A model-free machine-learning framework for solving optimal-control problems. It selects an action from a set of possible actions with a systematic method based on the maximisation of reward according to the behaviour of the mathematical model of a system. It is useful for the optimisation of long-term system performance.

**reinforcing feedback** → positive feedback

**relative degree** A difference between the degrees of the denominator and the numerator of a transfer function. **S**: pole excess

**relative encoder** → incremental encoder (1, 2)

**relative maximum** → local maximum

**relative minimum** → local minimum

**relative stability** A measure of the distance to the stability margin of a stable or an unstable system. It is usually described with the gain margin and the phase margin or with real parts of the roots of the characteristic equation.

**reliability** 1. A probability that a product, system or service performs its intended function adequately for a specified period of time in a defined environment without failure. 2. The consistency of a measuring method where the application of the same method to the same sample under the same conditions should give the same result.

**relief valve** → safety valve

**remote centre-compliance device** A passive mechanical device at the robot end-effector, which allows small translational and rotational displacements that make assembly operations easier.

**remote control** A system, which monitors and controls a machine or a process mostly wirelessly from a distance, using communication signals, e.g., monitoring material or energy consumption, controlling pumps, valves, conveyors.

**remotely-piloted aircraft** → unmanned air-vehicle

**remote telecontrol unit** → remote terminal unit

**remote telemetry unit** → remote terminal unit

**remote terminal unit** A multipurpose microprocessor-controlled device, which is used for remote monitoring and control of various complex systems, e.g., offshore oil platforms, pump stations, water supply, wastewater collection, environmental monitoring, electrical power transmission networks. It interfaces with distributed control systems and SCADA systems, using, e.g., radio communications, microwave communications, satellite communications. **S**: remote telecontrol unit, remote telemetry unit, RTU

**repeatability** **1**. The closeness of the agreement between the results of successive experiments on the same object, conducted in a short time, using the same method performed by the same operator on the same location under the same conditions using the same measuring instrument. **2**. The ability of a sensor to provide the same or similar results of successive measurements under the same circumstances. It is an indicator of the consistency of a particular sensor and a measure of the absolute difference between a pair of repeated test results.

**reproducibility** The closeness of the agreement between the results of experiments on the same or similar object using the same method, performed by different operators on different locations, using different measuring instruments. It indicates that the obtained results are not the artefact of a particular experimental setup or a particular research laboratory. Therefore, it enables independent repetitions of experiments and measurements. It is a measure of the ability to replicate the findings of others.

**rescue robotics** A subfield of robotics, which supports emergency services by providing real-time video and other sensory data about the situation in the case of a disaster. Often, autonomous mobile robots equipped with victim-detection sensors collaborate as a coordinated team on ground, air or water.

**reset control** → integral control

**reset controller** → integral controller

**reset rate** A coefficient, which is the reciprocal of integral time and indicates the response speed of the I term of the controller.

**reset time** → integral time

**reset windup** → integral windup

**resistance level sensor** A level sensor, which consists of a steel-base strip and a flexible precision-wound helix resistor in the outer jacket of the flexible protective material, acting as a pressure-receiving diaphragm, that is immersed in the liquid. Below the surface, the resistor is shortened due to the contact with the base strip, which enables continuous level measurements. It has no moving parts and can also be used in explosive liquids.

**resistance temperature detector** A temperature sensor, the resistance of which changes proportionally to the temperature change. The resistive element is usually a wire-wound element, which consists of mostly platinum, nickel or copper wire wrapped around a ceramic bobbin in a metal-alloy housing. Due to its accuracy, stability and linearity over a wide temperature range, it is used in, e.g., industrial application, automotive application, aerospace application, medical application. **S**: resistance thermometer, RTD

**resistance thermometer** $\rightarrow$ resistance temperature detector

**resistive humidity sensor** A humidity sensor, which exploits the conductivity dependence of nonmetallic conductors on their water contents. The conductivity between noble-metal electrodes changes when the measured hygroscopic material, e.g., special substrate, polyelectrolyte, conductive polymer, which is inserted between the electrodes, absorbs water. The measured quantity is proportional to the measured relative humidity. It is used in, e.g., industrial applications, pharmaceutical applications, residential applications, commercial applications.

**resolution** The smallest increment of the measured value that a measuring instrument can detect and display. It expresses the fineness, to which a measuring instrument can still be read and is often given as the number of bits.

**resolver** An angular displacement sensor, which measures the absolute angular position of the shaft by converting mechanical motion to an electrical signal. It is constructed as a two-phase electric motor, consisting of a transformer, which supplies the rotor with power brushlessly, and a stator where the induced voltage enables the determination of the angular displacement of the shaft. It has no electronic components and can operate in harsh environmental conditions.

**resonant frequency** The frequency of a harmonic input signal, at which the largest ratio between steady-state amplitudes of output signals and input signals is reached.

**resonant peak** The highest value of the amplitude response of a dynamic system in the Bode plot. It occurs at the resonant frequency as the consequence of a conjugate complex pair of poles in the open-loop transfer function.

**response** The output signal of a system, which is the consequence of either an input excitation or nonequilibrium initial state, e.g., time response, frequency response, natural response.

**response time** The property of a system, which defines the dynamics of the system response to the applied external excitations, described by transient-response specifications, e.g., delay time, rise time, settling time.

**restricted space** The subspace within the maximum space, to which a robot is restricted by limiting devices that reduce the range of motion and establish limits, which cannot be exceeded.

**retention valve** → check valve

**retro-reflective proximity sensor** A photoelectric proximity sensor or ultrasonic proximity sensor, which consists of an emitter that produces light or sound pulses, a receiver in the same module and a reflector that bounces the light beam or sound back to the receiver. The module and the reflector are installed in two opposing locations and an object, passing between the two, breaks the light beam or sound pulses causing the reaction of the sensor output.

**retrospective validity** A model-validation procedure, which determines the degree to which the model response fits the past data.

**reverse-acting control** → reverse action

**reverse acting control valve** A control valve, which tends to be closed when the flow rate increases. It requires a special construction of the valve actuator, as well as of the valve disc and the valve seat.

**reverse action** The mode of operation of an industrial controller in the case that negative gain is introduced by the actuator or by the process. Therefore, the control variable must be multiplied by $-1$ to achieve negative feedback. For instance, the increase of the output temperature of the system above the reference temperature requires the opening of the valve for the supply of the cooling medium. **S**: reverse-acting control

**reverse path** → feedback path

**revolution counter** → tachometer

**reward function** → fitness function

**R-factor** → thermal insulance

**rheometer** A viscometer, which measures visco-elastic properties of materials under various kinds of stress or strain, e.g., rotational rheometer, capillary rheometer, extensional rheometer. It can also measure the viscosity of non-Newtonian fluids, e.g., semi-solids, suspensions, emulsions, oil, asphalt, polymers, wax, paints, coatings, adhesives. It is used in material science as well as in e.g., chemical industry, pharmaceutical industry, cosmetic industry, food industry, beverage industry. **S**: plastometer

**Riccati equation** A first-order ordinary differential equation, which is quadratic in the unknown function. Its algebraic version is mostly given in the matrix form. It is used for optimal-control design or robust-control design, as well as for optimal-filtering design, all in the continuous-time domain or in the discrete-time domain.

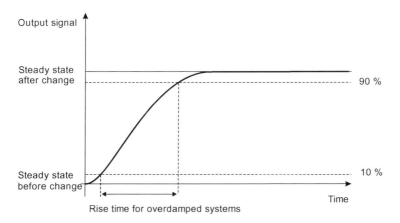

**Fig. 5** Rise time (1) for overdamped second-order systems

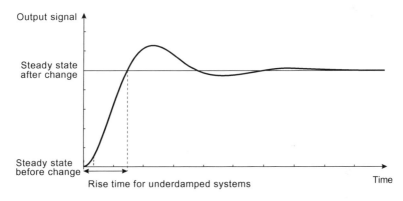

**Fig. 6** Rise time (1) for underdamped second-order systems

**rise time 1**. The transient-response specification defining the time required for the step response of a proportional system to rise from a minimal value to a value within a prespecified band around the steady-state value. The considered values are usually 10 % to 90 % for overdamped second-order systems and 0 % to 100 % for underdamped second-order systems. See Figs. 5 and 6. **S**: time-rises **2**. Parameter, defined by the difference between the intersections of the tangent at the inflexion point of the reaction curve with the steady-state lines before and after the step-input change. It is used for tuning a PID controller, e.g., with the step-response tuning method. See Fig. 7.

**RK method** ↔ Runge-Kutta method

**roboethics** → robot ethics

**robot** An intelligent moving mechanism, which is capable of carrying out the required tasks, e.g., robot manipulator, mobile robot.

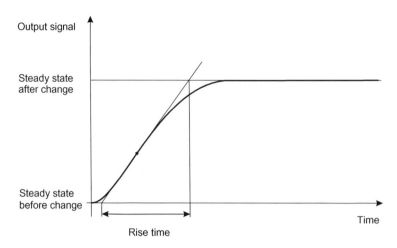

**Fig. 7** Rise time (2)

**robot animal**  A mechanical representation of a living creature, which mimics some animals, e.g., robotic snake, robotic insect, robotic fish. It is able to perform diverse tasks that may be difficult for a human. It can serve also as a robotic toy or pet. **S**: animal-inspired robot, animal robot

**robot arm**  A mechanism resembling a human arm, which can act independently or as a part of a more complex robot. It consists of a serial kinematic chain of rigid robot links, connected with joints and powered with an actuation system that provides positioning of the robot gripper in the reachable workspace.

**robot base**  A platform, to which the robot arm is attached. It is mounted at the end of a kinematic chain opposite to the robot end-effector.

**robot cell**  A group of robots, workstations, storage buffers, transport systems, part positioners, safety environment, and CNC machines, in which a single family of parts is produced. **S**: robot workcell

**robot end-effector**  A mechanism mounted at the end of a kinematic chain opposite to the robot base, such as a robot gripper or a tool, e.g., spraying nozzle, welding gun.

**robot ethics**  A set of rules, the objective of which is to develop scientific, cultural, and technical tools to promote and encourage the development of robotics for the advancements of human society and individuals, and to help prevent misuse of robotics against humans. **S**: roboethics

**robot gripper**  A device, which enables physical interaction between the robot and the workpiece, e.g., vacuum gripper, pneumatic gripper, hydraulic gripper, servo-electric gripper, magnetic gripper.

**robot hand guiding** A robot collaborative technique, in which a special guiding device located near the robot end-effector is used to transmit motion commands to the robot system. This allows unskilled users to interact and program the robot in a more intuitive way than using a teach pendant. It is suitable for applications, in which the robot system acts as a power amplifier.

**robotic assembly** Robot operation for putting together manufactured constituent parts to make a complete product. It saves workers from tedious and dull production line jobs and increases quality and throughput in, e.g., electronic industry, electromechanical industry.

**robotic coating** Robot operation for implementing material, e.g., paint, ink, varnish onto the surface of a workpiece. It results in reduced consumption of the material as well as in the higher quality of the finalised workpiece and reduced exposure of humans to toxic materials. **S**: robotic spray painting

**robotic die casting** Robot operation for handling parts in a machine that forces molten material under high pressure into a mould cavity. Therefore, hot, dirty and hazardous operations, which are unpleasant or dangerous for human workers are performed by a robot.

**robotic dispensing** Robot operation for moving various materials, e.g., fluids, adhesives, wire, medications, concrete, DNA genetic material, from one location to the other with high accuracy, precision and consistency. It results in material savings, in higher quality products, in complete inspection and in safety improvements.

**robotic exoskeleton 1**. A mechanism with rotational joints, which can be attached either to the entire human body or to the upper or lower extremities. It is mainly used for rehabilitation, as well as in teleoperation systems and for imitation learning of humanoid robots. **S**: powered exoskeleton, wearable robot **2**. A body-mounted unit working in a tandem with the user. It is powered by electric motors, pneumatic motors or hydraulic motors and equipped with the corresponding sensors and actuators. It can be either hard-type, e.g., powered armour, or soft-type, e.g., robotic suit, according to the used materials. It can be applied in, e.g., medicine, military, industry. **S**: powered exoskeleton, wearable robot

**robotic finishing** Robot operation for performing continuous-path movements needed for finalising tasks, e.g., robotic coating.

**robotic inspection** Robot operation for performing QA, QC and testing using robot manipulation and corresponding sensors, e.g., video system, laser sensor, ultrasonic sensor. It checks the compliance of a workpiece or assembly according to the specifications.

**robotic machine loading** Robot operation for grasping a workpiece from, e.g., conveyor belt, pallet, container, to orient it correctly and to insert it into a machine, e.g., lathe, CNC machine, plastic injection blow machine. After processing, the robot unloads the workpiece. The appropriate operation efficiency is achieved when a single robot is used to service several machines.

**robotic maintenance** Robot operation for keeping the condition of devices and for repairing devices in, e.g., nuclear industry, highways, railways, power-lines maintenance, aircraft servicing. Often, teleoperated and autonomous robots are used.

**robotic manipulation** Robotic handling of the object by moving, inserting or orienting it, to be in the proper pose, e.g., pick-and-place, multi-robot collaborative operation, as well as destroying the object. The constraints of the physical environment must be considered.

**robotic mapping** A procedure, which enables the mobile robot to acquire a spatial model of its physical environment. It constructs an indoor or outdoor map that is mostly used for navigation. The needed information is gathered from the sensors that perceive the outside world, e.g., digital camera, distance sensor, radar, tactile sensor, compass, GNSS.

**robotic material handling** Robot operation for transporting objects, in which the robot cooperates with material handling devices, e.g., containers, pallets, loading bins, conveyors, guided vehicles, carousels. It is used for performing the most tedious, dull and unsafe tasks in a production line.

**robotic micromanipulation** A technology, which enables the execution of a task with the microrobot system.

**robotics** An interdisciplinary field of engineering and science that uses mechatronics, artificial intelligence, control technologies, information technologies, computer technologies and other technologies to design, build and apply mechanisms that can replace human beings in the execution of physical activities and decision-making tasks. It deals with the intelligent movement of various robot mechanisms, e.g., robot manipulator, intelligent vehicle, collaborative robot, humanoid, biologically-inspired robot.

**robotic sealing** Robot operation for performing continuous movement along a predefined path while applying a precise amount of sealant by controlling different operation variables.

**robotic sorting** Robot operation, which, using the corresponding sensors, distinguishes different types of items classifying them into appropriate groups, e.g., cartons sorting, cards sorting, waste sorting.

**robotic spray painting** → robotic coating

**robotic welding** Robot operation for joining materials by using heat to melt and fuse the parts as well as for the handling of workpieces. It is used for, e.g., spot welding, arc welding, laser welding. It is the most frequent robot operation for improving speed, precision, and accuracy.

**robotic wrist** A mechanical system positioned between a robot arm and a robot gripper, usually with three rotational joints. Its axes intersect at the same point enabling the required orientation of the object grasped by the robot gripper.

**robot joint** A mechanism, which provides controlled relative motion between two robot links, e.g., prismatic joint, rotational joint, universal joint, spherical joint. Such mutually-attached bodies reduce the number of DOF to less than the number of DOF of a free body.

**robot language** A structured domain programming language, which incorporates high-level statements. It is an interpreter-oriented simulation language, in which an interactive environment allows the programmer to check the execution of each statement before proceeding to the next one. It enables online programming, using the actual robot in situ, as well as offline programming, using software tools without occupying the robot.

**robot learning** An operation enabling a robot to acquire novel skills or adapt to its environment using machine-learning algorithms, e.g., reinforcement learning.

**robot link** A basic part of a robot mechanism, which connects two neighbouring robot joints. **S**: robot segment

**robot machining** Robot operation for performing controlled material-removal processes, e. g., drilling, milling, grinding, routing, deburring. The robot holds either a workpiece or a powered tool.

**robot manipulator** A moving mechanism, which is designed to perform a task in space. It mostly consists of a robot arm, robot wrist, and robot gripper.

**robot mobility** The property of a kinematic structure, defined by the minimum number of independent robot-joint variables, which must be specified to enable the complete description of the pose of all robot links of the structure at a given time instant. **S**: degree of freedom (4)

**robot pushing** The motion of a robot or a mobile robot, which moves an object by applying external force with a robot gripper or a tool. The tool significantly extends the manipulation repertoire of the robot, e.g., a series of pushes with a flat fence can be used to eliminate the uncertainty in the orientation of a polygonal object.

**robot segment** → robot link

**robot simulation** Experimentation with a corresponding mathematical model, which enables off-line design, programming and validation of a robot system or a robot cell in a virtual environment.

**robot system** A hardware and software system, which includes a robot manipulator, a control system, a robot gripper, sensors, actuators, and sometimes a teach pendant and a safety environment, as well as a workpiece manipulated by the robot.

**robot-task frame** → world coordinate frame

**robot workcell** → robot cell

**robot workspace** The set of all positions that the robot end-effector can reach. **S**: effective workspace

**robustness 1.** The property of a closed-loop system to behave according to the selected criterion for all possible parameter values with all possible uncertainties. **2.** The property of the closed-loop system to behave within the specified limits regardless of the parameter-value change or the change of controlled-process structure. **3.** The property of a model, that is not too sensitive to the errors in the input data, e.g., measuring noise, missing data.

**robust performance** A closed-loop system behaviour, with the evaluated nominal performance and the guaranteed robust stability, which maintains the behaviour within acceptable limits in spite of uncertainties.

**robust stability** The stability of a closed-loop system, achieved with the controller designed for the nominal system, which is able to stabilise the system with structured and unstructured uncertainties.

**Rodrigues' formula** The equation, which describes rotation about an arbitrary axis.

**roll angle** The angle, which determines the rotation around the longitudinal axis of an object or a mobile system, e. g., aeroplane, ship, robot end-effector. Besides the yaw angle and pitch angle, it is an element from the set of three angles that completely determine the orientation of an object in space. **S**: bank

**roll-pitch sensor** → inclinometer

**roll, pitch, yaw angles** → Tait-Bryan angles

**root contours** A root locus analysis for different variable parameters, usually not considering the positive control-loop gain. However, only one parameter is treated at a time, e.g., a pole of the open-loop transfer function.

**root locus analysis** A graphical method for examining changes in the positions of the closed-loop poles in the $s$-plane or in the $z$-plane. The changes are the consequence of certain parameter variations, commonly positive control-loop gain. It is used for the investigation of the parameter influence on the closed-loop system behaviour, especially its stability, and for the control design.

**root locus diagram** → root locus plot

**root locus graph** → root locus plot

**root locus plot** A graphical representation of the values of complex variable $s$ in the roots of the characteristic equation, which satisfy both the magnitude condition and the angle condition. The roots change their positions under the influence of a certain parameter variation, commonly positive control-loop gain, enabling the investigation of closed-loop-control-system stability using the open-loop transfer function. **S**: root locus diagram, root locus graph

**Rosenbrock system matrix** A representation of a linear MIMO system with a polynomial matrix. It is obtained from the mathematical model of the system described with differential and algebraic equations after taking the Laplace transform with zero initial conditions. It bridges the gap between the state-space model and the transfer-function-matrix representation.

**rotameter** A flow meter, which measures the volumetric flow rate of liquid or gas on the variable area principle. A float rises in a vertical, slightly conically shaped tube under the influence of the flowstream. The displacement of the float is proportional to the measured flow rate. **S**: variable-area flowmeter

**rotary damper** An idealised lumped-parameter element for modelling a rotational mechanical system, which dissipates energy through rotary-motion damping. It has the property of resistance, e.g., toothed gear rotating in a viscous fluid. **S**: rotational damper

**rotary encoder** 1. An absolute encoder, which measures angular displacement or movement of a rotating shaft, generating an analogue signal or a digital signal, e.g., mechanical encoder, optical encoder, magnetic encoder. It is commonly used in, e.g., manufacturing robotics, medical industry, CNC machines. **S**: shaft encoder 2. An incremental encoder, which measures angular displacement or movement of a rotating shaft, generating an analogue signal or a digital signal, e.g., mechanical encoder, optical encoder, magnetic encoder. It is commonly used in, e.g., manufacturing robotics, medical industry, CNC machines. **S**: shaft encoder

**rotary motion sensor** → angular displacement sensor

**rotary potentiometer** A resistive angular displacement sensor, which measures the circular motion of an object connected to a sliding element of the voltage divider converting it to an electrical signal. The change in angle causes the resistance change between the fixed point and slider, which is proportional to the measured angular displacement.

**rotary speed** → angular velocity

**rotary torque sensor** A torque sensor, which converts dynamical torsional mechanical input into an electrical output signal. It requires the transfer of electric, magnetic or optic signal from a rotating system, typically shaft, to static electronics, using e.g., slip ring, rotary transformer, wireless telemetry.

**rotary-vane flow meter** A positive-displacement flow meter, which measures volumetric flow rate by creating a precise volume of fluid with each revolution of an impeller with two or more vanes. The impeller is driven by the flowstream of fluid pushing it to the output of a sensor, where each rotation is counted.

**rotary vane pump** A positive displacement pump, which transfers fluid using paddle wheels preventing fluid leakage from one chamber to another, e.g., sliding-vane pump, flexible-vane pump, swinging-vane pump, rolling-vane pump. It excels at handling low-viscosity liquids in various industrial, vacuum, pressure and combined applications, e.g., air conditioning, refrigeration, beverage industry, automotive industry.

**rotating paddle level switch** A level switch, which consists of an electric motor that rotates a shaft with mounted blades. When the measured medium covers the blades, the torque requirement increases and the motor revolutions decrease. This causes the corresponding motor switching and creates a signal for the control system. It is used for point level detection for powder or granular solid materials in applications dealing with, e.g., fertilizers, pesticides, chips, pharmaceuticals. **S**: paddlewheel level switch

**rotational damper** → rotary damper

**rotational joint** A single-DOF robot joint, which constrains the movement of two neighbouring robot links to angular displacement. The relative position of one robot link with respect to the other is given by the angle around the robot joint axis.

**rotational mechanical system** A system, which consists of idealised, weightless and dimensionless elements for circular motion, e.g., a moment of inertia, torsion spring, rotary damper.

**rotational motion** → circular motion

**rotational spring** → torsion spring

**rotational velocity** → angular velocity

**rotational viscometer** A viscometer, which consists of a turning spindle in a cup with sample material, e.g., Couette rotational viscometer, Searle rotational viscometer, Stabinger rotational viscometer, cone and plate rotational viscometer. The torque on the vertical shaft that revolves around the spindle or the cup is proportional to the measured viscosity. The torque is measured either through deflection of the spring that stops the revolving part or through servomotor current needed to maintain a certain angular velocity. Different shapes of spindles enable the adaptation of the device to the tested material.

**rotation matrix** An orthogonal 3x3 matrix, the elements of which are the direction cosines of the angles between individual axes of two coordinate frames. It can describe either the rotation or the orientation of a coordinate frame.

**roundoff error** An error of the numerical integration method, represented as the difference between the approximation, used in computation, and the exact value of the integral. It is a consequence of the numerical constraints of the digital computer performing numerical integration and depends on the length of the simulation run, on the complexity of the used integration algorithm as well as on the step size.

**Routh-Hurwitz stability criterion** A method for analysing the stability of a linear time-invariant closed-loop SISO continuous-time system. The stability depends on the number of poles in the right $s$-halfplane, which is determined using a table that is based on the coefficients of the characteristic equation. **S**: Routh stability criterion

**Routh stability criterion** → Routh-Hurwitz stability criterion

**RPM gauge** → tachometer

**RS 232** A standard for serial, long-distance bit-by-bit transmission of binary data between two devices, e.g., a computer and a peripheral device. An additional converter is often needed for the connection to a computer.

**RS 485** A standard for serial, long-distance, high-speed transmission of binary data between multiple devices, enabling the creation of a network. An additional converter is often needed for the connection to a computer.

**RTD** ↔ resistance temperature detector

**RTU** ↔ remote terminal unit

**Runge-Kutta method** A single-step integration method for approximate numerical solving of ordinary differential equations. For the given initial value and step size, the next values of the integral are in most cases calculated using the corresponding formula and the coefficient scheme. **S**: RK method

**R-value** → thermal insulance

**RW** ↔ reaction wheel

# Chapter 20
# S

**saddle point** An equilibrium point of a second-order system in the phase plane, where all trajectory derivatives equal 0. However, it is not a local extremum.

**safety valve** A valve, which prevents the pressure in a system to exceed the prescribed maximal operating pressure. It is usually counterbalanced by a dead-weight, by a spring or by a hydraulic assembly. **S**: relief valve

**sample** One of the consecutively measured values of a sampled signal in a particular time instant.

**sampled-data control** → discrete-time control

**sampled-data model** → discrete-time model

**sampled signal** A discrete-time signal representing a continuous-time signal. It is obtained by acquiring the continuous-time signal in consecutive time instants. Its amplitude can take any value from a continuous interval of possible values and is equal to the value of the original continuous-time signal in each time instant.

**sampled-time control** → discrete-time control

**sampler** A signal processing device, which is used in an A/D converter for converting a continuous-time signal to a discrete-time signal by acquiring the continuous-time signal in consecutive time instants. Its output is a sampled signal, which represents the original continuous-time signal. **S**: sampling element

**sampling** A process that acquires a continuous-time signal in consecutive discrete instants. It results in the sampled signal, which represents the original continuous-time signal.

**sampling controller** → discrete-time controller

**sampling element** → sampler

© ZRC SAZU/Research Centre of the Slovenian Academy of Sciences and Arts 2023
R. Karba et al., *Terminological Dictionary of Automatic Control, Systems and Robotics*,
Intelligent Systems, Control and Automation: Science and Engineering 104,
https://doi.org/10.1007/978-3-031-35755-8_20

**sampling frequency** → sampling rate

**sampling interval** → sampling time

**sampling period** → sampling time

**sampling rate** The number of samples of a continuous-time signal per unit of time, which is used for generating a discrete-time signal. **S**: sampling frequency

**sampling theorem** A theorem stating that a frequency-limited continuous-time signal can be unambiguously defined using its consecutive discrete-time signal values if the sampling rate is at least double the highest frequency in the spectral density of the original signal. In such a case, the original signal can be fully reconstructed from its consecutive discrete-time signal values using Whittaker-Shannon interpolation. **S**: Nyquist-Shannon theorem, Shannon theorem

**sampling time** The time that passes between two consecutive measurements of a continuous signal or two consecutive data records. It defines the frequency of data collection. **S**: sampling interval, sampling period

**saturation** **1**. A property of a nonlinear system, which limits a signal value to the interval between the prescribed upper limit and the prescribed lower limit. **2**. A nonlinear block of a simulation language, which limits a signal value to the interval between the prescribed upper limit and the prescribed lower limit.

**SCADA** ↔ supervisory control and data acquisition

**scale model** A physical model with a static character that provides accurate information on proportions, colour, surface structure and appearance of the modelled system, e.g., car, aeroplane, ship, house, traffic object.

**SCARA** ↔ selective compliance assembly robot arm

**SCARA robot** → selective compliance assembly robot arm

**schematic model** A symbolic model, which provides the visualisation of the modelled system structure and operation. It is given in the form of, e.g., flowchart, signal-flow graph, block diagram, plan, map.

**screw** A pose of a rigid body, which can be described by a translation along the line and a rotation about the same line. It is defined by a six-dimensional vector that consists of a pair of three-dimensional vectors, describing twist and wrench.

**SD** ↔ system dynamics

**s-domain** Analytic space, in which the set of values that are accepted as the function input is covered by complex variable $s$ as the independent variable. **S**: Laplace domain

**SEA** ↔ series elastic actuator

**secondary battery** → rechargeable battery

**secondary controller** → slave controller

**secondary control loop** → slave control loop

**second controller** → slave controller

**second method of Lyapunov** → Lyapunov's direct method

**second-order system** A dynamic system that is described with a second-order differential equation or a second-order difference equation. It can approximate many real-life systems.

**selective compliance assembly robot arm** A robot arm, which has two rotational joints and one prismatic joint. Its reachable workspace is cylindrical. Its speed and repeatability enable its use in diverse small robotic-assembly applications as well as in pick-and-place. **S**: SCARA, SCARA robot

**self-organisation** A fundamental principle of structure formation and growth, which enables individual components to coordinate at a local level to achieve a common goal. The system evolves dynamically in joining building blocks to develop complex functioning units. According to some criteria, they operate better, without any external control. It appears in, e.g., robotics, manufacturing control, sensor networks, multi-agent systems, computational grid, ANN.

**self-organising map** An ANN that is optimised with unsupervised learning and is intended for dimension reduction of the input-data space, e.g., Kohonen map.

**self-reconfigurable modular robot** A modular robot, which can change the shape of its unit aggregation by rearranging physical connections among the units. It is mostly an autonomous and heterogeneous kinematic machine that is coordinated and synchronised to perform a task.

**self-regulation** The inherent ability of a system to reach a new steady state without the intervention of a controller, either at a constant input signal or after a sustained disturbance. **S**: inherent regulation

**self-tuning** The procedure in which the controller parameters adapt to changes in process operation, optimising a given objective function. **S**: parameter adaptation

**self-tuning control** The control, which adapts the controller parameters according to changes in corresponding parameters of the mathematical model of a system. The model is obtained online using parameter estimation. **S**: model-identification adaptive control, parameter-adaptive control

**self-tuning controller** An adaptive controller, which continuously updates its parameters to maintain optimal closed-loop performance, according to the online estimated parameters of a controlled system. It simultaneously executes tuning and control functions. **S**: parameter-adaptive controller

**selsyn** → synchro

**sense-plan-act paradigm** Control methodology, which enables the design and execution of the movements of the robot. It utilises sensor data and internal environment model. **S**: SPA paradigm

**sensing element** A primary element of a measurement system, which is in direct contact with the measurand, converting it into a quantity more appropriate for further processing. **S**: sensor (3)

**sensitivity 1**. A measure of responsiveness to changes in a system or to changes at its input. In system theory, it is often expressed as the ratio between system response change and parameter change. **2**. The slope of the static characteristic of a measuring system. It is usually expressed as the ratio between the number of the instrument scale increments, i.e., the output quantity increment, and the unit of the measured quantity at the input.

**sensitivity function** The transfer function between the output disturbance and the controlled variable in a closed-loop system. It is used in the analysis of the control system for disturbance rejection.

**sensor 1**. A device, module or subsystem, which detects events or changes in the environment and converts the measured electrical or nonelectrical quantity into the corresponding signal that can be further processed, e.g., pressure sensor, flow meter, displacement sensor, proximity sensor. **2**. A compact measurement system that includes all the necessary transducers, transmitters and converters. **3**. → sensing element

**sensor fusion** A combination of data from several sensors or a combination of data from diverse sources, to assure resulting information with less uncertainty than the one obtained from individual sources. It ensures better design, control and operation of a system. It is used in, e.g., military application, industrial application, robotic application, automotive application, medical application, consumer application, image processing. **S**: sensor integration

**sensor integration** → sensor fusion

**sensorless brushless DC motor** A brushless DC motor, which computes the rotor position from current and voltage measurements on the stator windings. Therefore, it eliminates the need for measuring the angular displacement.

**sequence control** → sequential control

**sequential control** Logic control, which defines an ordered set of individual tasks. The binary outputs depend on the actual binary inputs, as well as on the previous signals and events. It is implemented using, e.g., memory cells, timers, counters. **S**: sequence control

**sequential function chart** A textual and graphical standard PLC programming language for structuring and organising programs and function blocks.

**sequential logic** A mathematical formalism, which describes sequential control that can be implemented using Boolean elements and memory cells.

**sequential process 1**. The process with a sequence of independent states. **2**. The execution of a sequence of work-specific phases. **S**: plan-driven process

**serial correlation** → autocorrelation

**serial manipulator** → serial robot

**serial robot** A robot, the robot base of which is connected to the robot end-effector by an open kinematic chain. Its structure is often anthropomorphic. It is the most common industrial robot. **S**: serial manipulator

**series compensation** Compensation, in which the compensator is connected to the input or to the output of the controlled system in the direct path of the control loop.

**series elastic actuator** A combination of a mechanic power source, e.g., motor gearbox, which is connected in series to an elastic element, e.g., spring. The twist of a spring is measured to control the force output. The measurement of the twist is thus used as a force sensor. It is used with collaborative robots. **S**: SEA

**service robotics** A subfield of robotics, in which nonindustrial robots perform useful tasks for humans or equipment. It is used in, e.g., healthcare, agriculture, logistics, defence, safety, cleaning, maintenance, food delivery, entertainment.

**servomotor** A rotary or linear electric motor, which allows precise control of angular or linear position, velocity and acceleration using position feedback. It is commonly used in, e.g., robotics, CNC machinery, automated manufacturing, process industry, home electronics, toy industry.

**servosystem** A feedback control system, the output of which is mechanical motion.

**setpoint** An independent value, which defines the wanted ideal value or the prescribed course of the controlled variable in a control system. It is usually set by an operator or by a supervisory control system. **S**: reference

**setpoint control** → tracking control

**setpoint tracking** Operation of a control system, which is primarily designed to ensure that the controlled variable follows the frequently changing setpoint signal as closely as possible. On the other hand, the eventual disturbances influencing the controlled variable are of secondary concern. **S**: reference tracking

**settling time** The transient-response specification defining the time, in which the response of a proportional underdamped second-order system to a unit-step signal enters and stays within a prespecified range around the steady-state value. The range is given as the percentage of the steady-state value. Frequently used values are 5 % or 2 %.

**shaft encoder** → rotary encoder (1, 2)

**Shannon theorem** → sampling theorem

**shift** → linear displacement

**shut-off valve** → on-off valve

**sight glass** A level sensor, which consists of a transparent tube connected to a vessel at the bottom and at the top. The level in a tube is the same as in the vessel enabling the observation of the liquid level contained within. Level monitoring is important in a variety of applications dealing with filling or storing liquids in a vessel.

**signal** 1. A physical quantity that conveys information, e.g., voltage, current, electromagnetic wave. It is time-dependent and position-dependent. 2. An element of a block diagram or of a signal-flow graph, which shows a path of the connection between blocks with a line and its direction with an arrow.

**signal conditioning** A signal processing, which is needed in the case of nonmatching signals in a control loop. It is carried out by a corresponding signal converter.

**signal converter** A converter, which transforms a signal from one form to another form without loss of information, e.g., current-to-pressure converter and vice versa, current-to-voltage converter and vice versa, voltage-to-frequency converter and vice versa, frequency-to-frequency converter, A/D converter and vice versa.

**signal-flow diagram** → signal-flow graph

**signal-flow graph** The graphical representation of a system structure and unidirectional signal connections, which consists of the weighted and directed branches connecting the nodes. It represents a set of simultaneous algebraic equations describing a system in a cause-and-effect way. **S**: signal-flow diagram

**signal-flow-graph algebra** The set of rules which enables the transformation of a signal-flow graph to an equivalent form, e.g., simplification of the signal-flow graph, structural changes of the signal-flow graph.

**signal-to-noise ratio** The ratio of the power of the information-carrying signal to the noise power. It is usually expressed in decibels. **S**: S/N, SNR

**silent wave motor** → ultrasonic motor

**silicon bandgap temperature sensor** → integrated-circuit temperature sensor

**silicon-controlled rectifier** $\rightarrow$ thyristor

**SIMO system** $\leftrightarrow$ single-input multiple-output system

**simulation** The experimentation with the corresponding mathematical model, enabling the study of the system behaviour in real time, shortened time or extended time. Various types of differential equations can be solved by consecutive integration using dedicated software.

**simulation diagram** $\rightarrow$ simulation scheme

**simulation environment** A modelling framework providing software, which facilitates the experimentation with mathematical models, e.g., MatLab/Simulink, Modelica. **S**: modelling environment

**simulation language** A digital simulation system, which describes the implementation of a simulation model on a computer, e.g., continuous simulation language, discrete-event simulation language. The model is defined by the user either in a relatively simple programming syntax or using a graphical preprocessor.

**simulation model** A mathematical representation of a system that is formulated in such a manner that it conforms to specific syntactic rules of a particular simulation language, e.g., is given in a standard form, does not comprise any algebraic loops, is numerically stable. Therefore, it can be explicitly implemented in a simulation language.

**simulation package** A digital simulation system, which consists of the main program and a library of subroutines. On one hand, the user must be familiar with the programming language in which the package is written. On the other hand, it provides a high degree of flexibility.

**simulation run** A digital simulation implemented between the defined initial time and final time. The time interval is chosen so that it comprises the investigated system behaviour, e.g., a transient response.

**simulation scheme** The graphical representation of a simulation program, which consists of blocks, representing mathematical operations and corresponding connections. The basic blocks are integrator, summator, gain block and function block. Such a representation is the basis for analogue simulation, as well as for programming with block-oriented simulation languages. **S**: simulation diagram

**simulator** **1**. Hardware or software equipment, which enables realistic imitation of operation and control of a system. It is used for training people to operate complex or risky systems, e.g., aircraft, spacecraft, nuclear plant. It is also used for medical education as well as in the entertainment industry and as the support to operators to act in unexpected or unusual situations. **S**: testbed **2**. A programming environment, which enables a computer to execute computer simulation.

**simultaneous localisation and mapping** An algorithm, which constructs or upgrades a map of an unknown environment using an appropriate autonomous mobile system, while simultaneously determines its location using the developed map. **S**: SLAM

**single-degree-of-freedom system** The system with only one changeable dimension, parameter or variable, e.g., motion in one spatial dimension, a system with one control input, feedback control with a controller in only one path of the signal-flow graph or block diagram of the closed-loop system. **S**: one-degree-of-freedom system

**single-input multiple-output system** A dynamic system, which has one input and several outputs. Such structure enables the development of the controllable canonical form and thus facilitates controller design. **S**: SIMO system

**single-input single-output system** A dynamic system, which has one input and one output. It is often used in analysis and control-system design. **S**: SISO system, univariable system

**single-step integration method** A numerical integration method, which requires the value of the integral in only one previous step to calculate the value of the integral in the current step, e.g., Euler method, RK method, Gauss method, Milne method, Lobatto method, Rosenbrock method. **S**: one-step integration method

**singular value** **1**. The positive square root of an eigenvalue of the product of the square matrix and its transpose. **2**. The acceptable generalisation of SISO system gain, which offers a base for the MIMO-system performance assessment. It can be calculated from the transfer function matrix. **S**: principal gain, principal value

**singular-value decomposition** A matrix fraction method, which transforms a matrix into a product of a unitary matrix, a diagonal matrix and the unitary-matrix transpose. The diagonal matrix has nonnegative singular values arranged in descending order in its diagonal, while the unitary matrices are composed of input and output singular vectors, respectively. It can be used, e.g., for the matrix rank determination, in principal components analysis, in image processing.

**SISO system** ↔ single-input single-output system

**SLAM** ↔ simultaneous localisation and mapping

**slave controller** A controller in a cascade control structure, which uses the difference between the output of the master controller and the measured auxiliary controlled variable to create the auxiliary output, which is the control variable for the part of the process not included in the slave control loop. **S**: secondary controller, second controller

**slave control loop** The inner loop in a cascade control structure, which contains a slave controller and the part of the process affected by the slave controller. **S**: secondary control loop

**sliding dot product** → cross-correlation

**sliding inner product** → cross-correlation

**sliding-mode control** A variable-structure control, which tends to drive system states towards a specific surface in the state space. The discontinuous control law keeps the states in the near neighbourhood of this sliding surface. It offers robust control for complex, high-order nonlinear dynamic systems that operate under uncertain conditions, e.g., robot control, electrical-drive control, process control, vehicle control. It exhibits low sensitivity to parameter variations, which eliminate the need for exact modelling and enables easy tuning and implementation.

**slip sensor** A sensor, which measures the relative movement of one object surface over the surface of another object when they are in contact, e.g., distribution and amount of the tangential component of the contact force in a robot gripper.

**sluice valve** → gate valve

**small signal approximation** → linear approximation

**smart building** A building that is controlled and supervised using a BMS. S: intelligent building

**smart factory** An agile production facility, which is highly digitalised, connected and optimised. It drives a greater value from, within and across the supply network, customers and other factories. It enables predictive operation and maintenance, self-adaptation to changing conditions and needs, self-optimisation, self-learning, real-time or near-to-real-time reactions and autonomous running of the production process. It integrates different systems and methodologies, e.g., cyber-physical systems, control systems, robotic systems, IIoT, artificial intelligence, machine learning, HMI, cloud computing, cognitive computing.

**smart home** A building used for residential purposes, e.g., house, block of flats, which is controlled and supervised using a specially designed BMS. The system allows adjusting, e.g., temperature, lights, shades, as well as setting particular preprogrammed scenes for, e.g., working, relaxing, sleeping, reading, listening to music, watching TV.

**smart sensor** A sensor, which processes the data from the environment using a built-in processor to perform predefined functions. It is an integral element of the IIoT as well as of wireless sensor networks. It is used in monitoring and control of, e.g., smart grids, battlefield reconnaissance, exploration.

**Smith-McMillan canonical form** A TFM that is transformed into a diagonal form using elementary row and column operations. It defines the poles and the transmission zeros of a MIMO system. S: McMillan form

**Smith predictor** A controller for systems with dead time, which is comprised of an inner and an outer control loop. The inner control loop uses the model of the system without the dead time. The outer control loop considers the output disturbance by comparing the controlled variable and the delayed output of the system without dead time.

**S/N** ↔ signal-to-noise ratio

**SNR** ↔ signal-to-noise ratio

**sociable robot** → social robot

**social robot** An autonomous robot that interacts and communicates with humans and other autonomous physical agents in an acceptable manner, often to achieve emotional goals in, e.g., education, healthcare, quality of life, entertainment, communication, collaboration. **S**: sociable robot

**soft computing** A subfield of computer science, which aims to exploit tolerance for imprecision, uncertainty, approximate reasoning and partial truth to come closer to humanlike decision-making. It merges various computational methods and technologies, e.g., fuzzy logic, ANN, evolutionary computation, expert systems. It enables the design and use of intelligent systems in, e.g., control systems, robotics, image processing, smart instrumentation, biomedical applications, healthcare, home appliances. **S**: computational intelligence

**soft-finger contact** The contact between a robot and an object with two degrees of freedom. The rotation about the contact normal is not possible because of the friction between the finger and contact surface.

**softPLC** → programmable automation controller

**soft robotics** A subfield of robotics, which deals with robots that are constructed from highly compliant materials, similar to those found in living organisms. It enables humans and robots to interact and collaborate in industrial environments, in service, and in everyday settings. The safety of human-robot interaction must be ensured.

**soft sensor** A computer program for assessing the value of a nonmeasured variable. It uses several measurements of various interdependent variables as well as the model of the process in order to calculate the value of the nonmeasured variable by implementing, e.g., Kalman filter, ANN, fuzzy logic. **S**: virtual sensor

**soft valve seat** A valve seat made of plastic material, which completely prevents leakage. It is used in, e.g., ball valve, butterfly valve, diaphragm valve.

**solenoid** An actuator, which uses an electromagnet to generate linear displacement. A current signal in the coil causes the ferromagnetic core to be pushed or pulled towards the centre of the coil. When the excitation stops, the corresponding spring returns the ferromagnetic core to the starting position. It is commonly used in, e.g., valve, robot, actuator system, air conditioning system.

**solid-state temperature sensor** → integrated-circuit temperature sensor

**sonar** 1. A sensor, which measures the distance to objects on or under the water surface using sound propagation and reflection. It applies infrasonic as well as ultrasonic acoustic frequencies. **2.** → ultrasonic sensor

**sorting algorithm** 1. An algorithm that rearranges elements of a list in a certain order according to a comparison operator. **2.** An algorithm, which puts blocks in a block-oriented simulation language or equations in equation-oriented simulation language in such order that the input of a certain block or the right side of an equation is calculated before solving the equation, which describes the output of this block or the left side of this equation. It enables the correct execution of the simulation.

**space robotics** A subfield of robotics, which deals with robots performing mobility, manipulation, assembly or servicing functions in extreme environment considering special constraints of space missions. They assist astronauts or extend the areas and abilities of exploration on remote planets as surrogates for human explorers.

**SPA paradigm** ↔ sense-plan-act paradigm

**special-purpose simulation language** A simulation language, which is intended for the use in particular domains, e.g., electronics, biopharmacy, ecology. The corresponding adjustments to the specifics of the certain target domains greatly facilitate modelling and simulation, especially for the domain experts. **S**: problem-oriented simulation language

**specific heat capacity** The parameter of a mathematical model, describing the property of homogenous material to be able to store heat, defined as the ratio of the stored heat change to temperature change divided by the mass of the material sample. It is a measure of the ability of the material to store heat.

**specific thermal conductance** → thermal conductivity

**specific thermal resistance** → thermal resistivity

**spectral canonical form** → diagonal canonical form

**spectral density** Distribution of the amplitudes and the phase shifts of each frequency component of the signal as a function of the frequency. The integral of the function along an arbitrary frequency interval is a constituent part of the whole variance of the signal. In signal analysis, power spectral density and energy spectral density are often considered. **S**: frequency spectrum

**spectrophotometer** An optical analytical instrument, which measures the intensity of absorbed light in relation to the wavelength after it passes through sample solution, e.g., visible light spectrophotometer, ultraviolet spectrophotometer, infrared light spectrophotometer, atomic absorption spectrophotometer. Every compound absorbs, transmits or reflects light over a certain range of wavelengths enabling the determination of the components in a sample. The intensity of light with a certain range of wavelengths is proportional to the concentration of the individual component in the sample. It is used in, e.g., production facility, analytical laboratory, chemical engineering.

**speed of response** A dynamic property of a measurement system describing the rapidity with which a measuring instrument responds to the changes of the measured quantity.

**spherical joint** A three-DOF robot joint, which allows rotational movement around three perpendicular axes. It can be, e.g., a passive joint in a parallel robot.

**spherical robot 1.** A mobile robot, which has a shell that is shaped like a ball and contains an internal driving mechanism that enables it to move on the ground by rolling. The communication between the robot and the external control unit is mostly wireless. It is often used in, e.g., environmental monitoring, patrol, rehabilitation. **S:** ball-shaped robot **2.** → polar robot

**spiral point** → focus point

**s-plane** The analytic space of complex variable $s$, which is used for the presentation of, e.g., poles and zeros of a continuous-time-system transfer function. Its $x$-axis depicts real parts and the $y$-axis depicts imaginary parts of complex values of $s$. It is a graphical analysis tool in engineering and physics.

**spool valve** A pneumatic or hydraulic device, which controls the flow direction from the power source by combining or switching the path through which the fluid can travel between inlet and outlet ports. The fluid is diverted in the required direction with small displacements of two or more small pistons in a cylinder. A variety of combinations concerning the number of pistons and ports enables the adaptation to numerous applications. **S:** directional control valve

**square MIMO system** A MIMO system with an equal number of inputs and outputs described with a square TFM.

**stabilisability** The property of a system that has all the unstable states controllable and all the uncontrollable states stable. It is a weaker property than controllability. The stability of the system can be ensured despite some uncontrollable states.

**stability domain** The part of the analytic space used for feedback control design, e.g., $s$-domain, $z$-domain, frequency domain, in which design parameters, e.g., controller coefficients, yield an asymptotically stable closed-loop system.

**stability limit** → stability margin

**stability margin** The value of a parameter, state or input of a system, at which a stable system becomes unstable or vice versa. **S:** stability limit

**staircase signal** A signal, which consists of equally or differently sized time-delayed step signals.

**standard form** $\rightarrow$ canonical form

**standard PLC programming language** A set of rules, which is defined in the international standard IEC 61131-3 and is used for converting textual strings or graphical elements to generate machine code that can be implemented in a PLC, i.e., structured text, instruction list, ladder diagram, function-block diagram, and sequential function chart.

**star tracker** An optical navigation sensor, which compares the acquired image of the stars to their known absolute positions from a star catalogue to determine the orientation in space.

**state controller** A controller, the inputs of which are the actual states of the controlled system. The states are used as feedback signals in the control loop.

**state equation** A part of the mathematical model of a lumped-parameter dynamic system in state space. It defines mutual relations among the state variables as well as relations between the system inputs and the state variables. A model is described by a system of first-order ordinary differential equations or first-order ordinary difference equations, usually in the vector-matrix form.

**state estimation** A procedure for the determination of the value of system states, which uses a state observer to assess the values of the nonmeasured states. It is often used in the case when a state controller is implemented.

**state matrix** $\rightarrow$ system matrix

**statement list** $\rightarrow$ instruction list

**state observer** A system for the estimation of the value of system states, which gathers information from measured signals, usually the input signal and the output signal. It is often used in the case when a state controller is implemented.

**state space** A multidimensional Euclidean space with state variables on coordinate axes.

**state-space model** 1. A mathematical description of a parametric model of a higher-order dynamic system represented with a system of first-order differential equations or with a system of first-order difference equations. The corresponding state variables enable the complete determination of the future behaviour of the system. **S**: state-space representation 2. A mathematical model, which connects the states with the inputs and the outputs of a system.

**state-space representation** $\rightarrow$ state-space model (1)

**state-transition matrix** An exponential matrix $e^{At}$, where A denotes the system matrix and t denotes time. It is used in solving state equations as well as in the analyses of linear state-space models.

**state variable** **1**. An element from the set of variables in state space that completely determines the future system behaviour. It is not necessarily measurable or observable nor it always has a physical background. **2**. $\rightarrow$ stock

**state vector** A vector, which consists of all state variables. In the case of only one state variable it is reduced to a scalar form.

**static biped walking** Any stable biped walking pattern where the vertical projection of COM always stays inside the support polygon. This type of walking typically requires large feet and strong ankle joints.

**static characteristic** The relationship between two or more quantities of a system or of an instrument under static or stable conditions, e.g., relationship between the output and the input of a system in a steady state. It is usually given in a graphical form.

**static gain** $\rightarrow$ steady-state gain

**static model** **1**. A mathematical model, which describes the behaviour of the modelled system in steady state, e.g., with algebraic equations. It contains no internal history about previously applied input, internal variables or output. **2**. A physical model with static character representing the object in the reduced or in the increased scale, e.g., scale model, imitation model.

**static system** A system, the output of which at any time instant depends only on the input at the same time instant.

**stationary Kalman filter** An algorithm for calculating optimal estimates of the states of a noise-affected linear dynamic system in a steady state.

**steady state** The state of a system, which occurs after the end of the transient response.

**steady-state deviation** $\rightarrow$ steady-state error

**steady-state error** The difference between the controlled signal and the reference signal of a feedback control system after the transient response ends. **S**: offset error, steady-state deviation

**steady-state gain** The ratio between the output-signal change and the input-signal change between two steady states of a linear system or a nonlinear system. It can be experimentally determined either by the measurement of gain at a frequency of 0 Hz or by the measurement of the steady state of a stable system response to a unit step signal. **S**: DC gain, static gain

**step function** A function, which enables modelling of a quickly and abruptly changing test signal in dynamic-system analyses. Its value is 0 at negative values of the independent variable, and nonzero constant at positive values of the independent variable.

**step invariance** A time-response fitting method for the system discretisation, which results in a discrete-time model that has the same step response in the sampling instants as the original continuous-time model. **S**: ZOH equivalent

**step motor** → stepper motor

**stepper motor** A brushless motor, which converts digital pulses into precise angular movements of the shaft, e.g., permanent-magnet stepper motor, variable reluctance stepper motor, hybrid synchronous stepper motor. It is open-loop controlled and stops and stays at the reached prescribed position. It is commonly used in, e.g., robotics, CNC machinery, pick-and-place, consumer electronics. **S**: step motor

**step response** A time response of a system to a step signal. It gives information on how the system reacts to a sudden change of the input and how the system reaches a steady state when starting from some other steady state.

**step-response method** A method for tuning PID-type controllers using, e.g., Ziegler-Nichols methods, Chien-Chrones-Reswick method, Cohen-Coon method. The dead time, the rise time, the reaction rate or the time constant can be estimated from the reaction curve of an at-least-second-order proportional over-damped system. The estimated values are used to calculate the parameters of the controller. **S**: reaction-curve method

**step signal** An aperiodic standard signal for testing the behaviour of dynamic systems, which is modelled by a step function. It is commonly used as the basis for obtaining time specifications and step response of a dynamic system.

**step size** A constant or variable increment of the independent variable. In the digital simulation of dynamic systems, the latter is time. **S**: calculation interval

**Stewart-Gough platform** A parallel robot, the mobile platform of which is controlled by six actuated leg-shaped mechanisms. By shortening or expanding them, the platform can be moved into the desired pose.

**stiff integration method** A numerical integration method, which increases the accuracy and calculation speed in stiff-system simulation by the appropriate handling of the step size, e.g., Gear method, Rosenbrock-Wanner method.

**stiffness** A property of an object, reciprocal of flexibility, which causes an elastic body to resist deformation or deflection as the response to the applied force. The typical elements are elastic bodies, e.g., spring, torsion spring.

**stiffness control** Impedance control, in which only the static relationship between the robot end-effector pose deviation from the desired motion and the contact force or contact torque is considered.

**stiff system** A system, the responses of which contain both very fast and very slow dynamics when compared to each other. Its mathematical model has the ratio between the maximal value and the minimal value of the real part of the Jacobian matrix eigenvalues greater than 100. Consequently, its time constants are in very different size classes, which may cause problems in simulation and control.

**stochastic model** A mathematical model, which uses ranges of values of each variable represented as probability distributions. Due to the random character of the variables, the relations among them are described with probabilistic laws. Changes in its discrete-time output responses in each time instant are described by the samples of past responses and probabilities for their changes in future. **S**: probabilistic model (1)

**stochastic optimisation method** An optimisation method that generates and uses stochastic variables, e.g., genetic algorithm, particle-swarm optimisation, differential evolution.

**stochastic signal** $\rightarrow$ noise (1)

**stochastic system** A system, which exhibits nonnegligible random behaviour. Therefore, the same input does not always result in the same output. It is modelled by a stochastic model. **S**: probabilistic system

**stock** An element of a stock and flow diagram that denotes a state of a dynamic system. It accumulates the difference between incoming and outgoing flows and is usually depicted as a rectangle. **S**: accumulation, level, state variable (2)

**stock and flow diagram** A graphic model that enables the simulation of the modelled system using appropriate computer tools for system dynamics. It uses stocks, flows, links and dynamic variables to model the dynamic properties of the modelled system.

**straight-line Bode plot** $\rightarrow$ asymptotic Bode plot

**strain gauge** A sensor, which measures extremely small displacements caused by, e.g., pressure, force, weight, tension, heat, that change the resistance of the zig-zag path of parallel metallic or semiconductor wire or stripes, e.g., metal foil strain gauge, semiconductor strain gauge, thin-film strain gauge. The expansion or contraction of the pattern glued on a flexible backing causes structural changes in the material of stripes, an increase of the length and a decrease of the diameter or vice versa to achieve resistance change, which is proportional to the measured displacement. It is frequently used as one of the transducers in different measurement systems.

**Strecker stability criterion** $\rightarrow$ Nyquist stability criterion

**strictly proper transfer function** A proper transfer function, in which the denominator degree is greater than the numerator degree. It describes a causal system, with a relative degree greater than 0. Its response approaches 0 as the frequency approaches infinity.

**structural stability** 1. The property of a dynamic system that small perturbations do not affect its qualitative behaviour. 2. The property of a dynamic system, which is closed-loop stable regardless of the controller gain value.

**structural uncertainty** An uncertainty, which is the consequence of the differences between the system structure and its mathematical-model structure, caused by, e.g., linearisation, unmodelled dynamics, disregard of the delays in the system.

**structural validity** A model-validation procedure, which asseses if the model satisfactorily reflects the internal structure of the system.

**structured control language** $\rightarrow$ structured text

**structured environment** A robot workspace, which is clearly and meticulously defined and is therefore predictable.

**structured text** A textual standard PLC programming language that implements a program as an ordered set of high-level commands. **S**: structured control language

**structured uncertainty** A parametric uncertainty, which is expressed as the difference between the system behaviour and mathematical-model behaviour, where the spot in the structure at which the uncertainty occurs is known.

**structure identification** A part of the system identification that determines the structure of the mathematical model, e.g., determination of the model order, selection of the number of basis functions.

**subject-matter expert** $\rightarrow$ domain expert

**Sugeno fuzzy model** $\rightarrow$ Takagi-Sugeno fuzzy model

**summator** The block of a simulation scheme, which allows an arbitrary number of signals to be added or subtracted. **S**: adder

**summing junction** $\rightarrow$ summing point

**summing point** The element of a block diagram, depicted as a circle, which indicates the point where two signals are summed or subtracted. **S**: summing junction

**superposition principle** A property of a linear system that the response caused by two or more inputs and nonzero initial conditions is the sum of responses to the individual inputs and nonzero initial conditions obtained separately. **S**: superposition property, superposition theorem

**superposition property** → superposition principle

**superposition theorem** → superposition principle

**supervised learning** A family of machine learning methods used for determining the function from a set of functions, which best describes the mapping from input data to output data. It is used in, e.g., identification, function approximation, pattern recognition.

**supervisory control 1.** The part of procedural control, which integrates the operation among lower-level controllers and indirectly influences the controlled system by adjusting the setpoints of the lower-level controllers. **2.** The operation mode of procedural control, which monitors the operation of the system and takes action only when a possibility of undesired behaviour arises.

**supervisory control and data acquisition** An industrial computer-based system for supervision and control of technological processes. It supports real-time data collection and logging, as well as updates the automatically established values of the variables and parameters of the process. It acts as an HMI and enables the operator to directly interact with devices to check and change process variables and parameters. It can display errors and sound alarms, as well as accept the relevant operator's actions. **S**: SCADA

**supply chain** A network between a company and its suppliers to produce and distribute a specific product or service from the original state to the end-user. It includes various activities, people, resources, information and entities, e.g., producers, vendors, warehouses, transportation companies, distribution centres, retailers.

**support polygon** The contact area between the robot and the ground in humanoid walking. It is the convex hull of all points in contact with the ground for the phase when only one sole is in contact with the ground and when both soles are in contact with the ground.

**surgical robotics** A subfield of service robotics, in which robots participate in the planning and the execution of endoscopic and minimally invasive surgical procedures, providing high accuracy and repeatability. It is used in, e.g., preoperative planning, preoperative warm-up, surgical rehearsal, remote telesurgery, intraoperative navigation, telementoring.

**switched system** A hybrid system with both continuous states and discrete states, in which a change in the discrete states causes only a change in the dynamics of the system, while it does not cause any discontinuity in the continuous states.

**symbolic model** An abstract model, which is given verbally, graphically, schematically or mathematically.

**synchro** An angular displacement sensor, which measures the absolute angular position of the shaft by converting mechanical motion to an electrical signal. It is constructed as a three-phase electrical motor, consisting of an AC-excited rotor and a stator, where the induced voltage enables the determination of the angular displacement of the shaft. It has no electronic components and can operate in harsh environmental conditions, especially for brushless units. When paired up or chained up, it can be used for remote angle transmission or signal transmission. **S**: selsyn

**synchronous motor** An AC motor, in which the magnetic rotor rotates with equal speed as the rotating magnetic field on the stator that, in turn, depends on the frequency of the supplied current, e.g., permanent-magnet motor, reluctance motor, hysteresis motor. It is not self-starting and its speed is independent of the load. It is commonly used as an electromechanical energy converter, in precise positioning and precise constant speed applications, e.g., robot actuator, timing machine, servomotor, actuator system.

**system** An interconnection of independent but interrelated elements comprising a unified entity, which functions towards a common goal. The elements can vary a lot and they may be, e.g., devices, substances, hardware, software, equations, organs, methods, rules, procedures.

**system-development life cycle** A planning procedure for the developers, which is divided into several steps, from specifying, planning, defining, developing, testing, analysing, validating, producing, selling, delivering to start-up procedure, deployment, usage, maintenance, support, phase-out, retirement, recycling and disposal of the product. It helps to improve the quality and reliability of products, reduces time-to-market, reduces prototyping costs and maximises supply-chain collaboration. **S**: application-development life cycle, product life cycle

**system discretisation** A conversion of a continuous-time model of a dynamic system to a discrete-time model, which depends on the assumptions of the system behaviour between consecutive time samples. The assumptions dictate the choice of the method, e.g., frequency-response fitting, time-response fitting, hold equivalent, zero-pole mapping.

**system dynamics** A method for modelling primarily complex nontechnical systems, which implements several steps to make the process and the model more comprehensible. First, a CLD is used to describe the dynamic properties of the treated system. Next, a stock and flow diagram is developed. **S**: SD

**system identification** A methodology for building a mathematical model of a dynamic system using measurements of its input signals and output signals. When the same input signals are applied, the resulting responses of the model should be as similar as possible to the measured output signals. **S**: experimental modelling

**system matrix** A square matrix in the vector-matrix form of a linear state equation, usually marked with letter $A$, which defines the mutual relations among the state variables. **S**: state matrix

**system order** A property of a dynamic system, which is determined either by the order of a differential or difference equation in the mathematical model of the system, or by the highest power of the complex variable $s$ or variable $z$ in the denominator of the transfer function of the system, or by the number of state variables in the state-space model of the system.

**systems approach** A method that tends to find the most acceptable solution to a considered problem from the aspect of its integrity, environment and limitations.

**systems engineering** A transdisciplinary and integrative technology, which enables the successful implementation of complex systems, considering the whole system development life cycle. It tends to understand and enable the development, testing and deployment of the solutions to meet the needs, using scientific, technological and management methods. It overlaps several disciplines, e.g., control engineering, aerospace engineering, manufacturing engineering, software engineering, project management.

**systems integrator** A group of experts or an engineering enterprise that is capable to connect components and subsystems into a complete system, e.g., a complex control system. It enables coordinated action of hardware and software equipment of different vendors, collected through procurement and technical activities.

**systems theory** An interdisciplinary methodological science, which tends to combine the existing knowledge about different systems into a uniform approach. It bases on the properties common to the majority of systems, e.g., aims, states, limitations, stability, behaviour, control.

**system type** A property of a dynamic system, which is expressed with the number of poles in the origin of the $s$-plane. It is visible when the mathematical model of the system is given in the zero-pole-gain transfer-function form. **S**: type number

**system zero** The zero from the set of transmission zeros and decoupling zeros of a linear MIMO system. In the case of the minimal realisation it is equal to the transmission zero.

# Chapter 21
# T

**tachometer** A sensor, which measures the angular velocity of the rotating shaft or disc of a motor or a machine. Contact type device uses an optical encoder or a magnetic encoder, while noncontact type device uses a laser or infrared light, e.g., optical tachometer. It is commonly used in, e.g., automotive industry, marine industry, aviation industry as well as in machining and mechanical systems. It often displays the measured value in revolutions per minute. **S**: revolution counter, RPM gauge

**tactile sensor** **1**. A sensor, which models the sense of human touch, i.e., cutaneous sense and proprioception. It measures information from physical interaction with the environment. It is widely used in, e.g., robotics, computer hardware, security systems, touch-screen devices, tactile imaging. **2**. A sensor, which detects or measures the spatial distribution of contact force perpendicular to a predetermined sensory area, e.g., deformation-based sensor, resistive sensor, capacitive sensor, optical sensor, piezoelectric transducer, magnetic sensor, mechanical sensor. It is often applied in collaborative robots, e.g., contact-force distribution between robot fingers and a manipulated object.

**Tait-Bryan angles** Euler angles, which determine the orientation of an object in space with regard to three sequential rotations around three different axes of the coordinate system. The sequence of rotations around axes $x$, $y$ and $z$ can be one of the following: $x$-$y$-$z$, $y$-$z$-$x$, $z$-$x$-$y$, $x$-$z$-$y$, $z$-$y$-$x$ or $y$-$x$-$z$. **S**: Cardan angles, heading, elevation, bank, nautical angles, roll, pitch, yaw angles

**Takagi-Sugeno fuzzy model** A fuzzy model, the consequent part of which is a mostly affine crisp function of input variables, while the antecedent part divides the input space into fuzzy regions. No defuzzification is needed. **S**: Sugeno fuzzy model, Takagi-Sugeno-Kang fuzzy model

**Takagi-Sugeno-Kang fuzzy model** $\rightarrow$ Takagi-Sugeno fuzzy model

**takeoff point** $\rightarrow$ branch point

© ZRC SAZU/Research Centre of the Slovenian Academy of Sciences and Arts 2023
R. Karba et al., *Terminological Dictionary of Automatic Control, Systems and Robotics*,
Intelligent Systems, Control and Automation: Science and Engineering 104,
https://doi.org/10.1007/978-3-031-35755-8_21

214                                                                    21  T

**tangent-line approximation** → linear approximation

**target flow meter** A flow meter, which consists of a flat disc or sphere with an extension rod suspended in the flowstream. It measures the drag force caused by the flow, enabling the calculation of the measured flow rate. It can measure various liquids, steams and gases, including sporadic or multidirectional flows.

**Taylor-series expansion method** A linearisation, which gives the model of the considered system in the analytic form. The treated model is usually approximated with the zero-order term and the first-order term of the Taylor series to obtain a linear deviation model. Such an approximation is justified only in the previously specified vicinity of the operating point. **S**: analytical linearisation, Taylor-series linearisation method

**Taylor-series linearisation method** → Taylor-series expansion method

**teach box** → teach pendant

**teaching by showing** A process, in which the operator guides the robot along the desired motion path manually or using a teach pendant. During motion execution, the data read by the robot joint position transducers are stored and later used as a reference for the robot joint actuators.

**teach pendant** A portable hand-held device, containing pushbuttons, switches, joysticks or a touchscreen tablet, which may be wired or wireless. It is used for online programming and positioning of the robot end-effector. **S**: teach box

**technocentric approach** The use of hierarchical work organisation in the process of control systems design, enabling cost reduction and exclusion of human errors.

**technology** The application of scientific knowledge, which comprises techniques, skills, methods, tools, devices and processes. It is used in the production of goods and in the development of services, enabling the transformation of a system from the current to the desired form or behaviour. It is implemented for achieving some specific practical purpose in industry, as well as in everyday life.

**telemanipulator** A robot, which is remotely controlled by a human operator to bridge the gap between the operator and the working environment. The gap can be either physical distance in, e.g., space applications, marine applications, telepresence, or danger in, e.g., nuclear material handling, telemedicine. **S**: teleoperator

**teleoperation system** A system, which enables a human operator to remotely explore and manipulate objects. It consists of a master device, e.g., haptic interface, a slave device, e.g., telemanipulator that manipulates objects, and a controller that couples the master device and the slave device.

**teleoperator** → telemanipulator

**telepresence** The use of a virtual environment, which enables the operator to feel as being physically present at a remote site.

**temperature gauge** → temperature sensor

**temperature probe** → temperature sensor

**temperature sensor** A sensor, which measures the amount of heat stored in an object or in fluid, e.g., thermocouple, thermistor, RTD, liquid thermometer, bimetal thermometer, pyrometer. Under the influence of heat variations, corresponding changes are generated and converted to a numerical value. It can base on mechanical, electrical or radiation principles. The heat is transferred by conduction, convection or radiation. The measured value is expressed in kelvins, degrees Celsius or degrees Fahrenheit. It is widely used in, e.g., industrial applications, laboratory settings, home appliances. **S**: temperature gauge, temperature probe, thermometer

**temperature switch** A device, which senses temperature and changes its binary output at the moment its preset value is reached, e.g., mechanic temperature switch with bimetallic strip, electronic temperature switch with a thermistor. It is commonly used in general industrial applications, as well as in power supplies, heating systems, cooling systems, household systems. **S**: thermal switch

**tendon drive** A transmission system, which connects a motor and remote mechanism via flexible cables and pulleys.

**testbed** → simulator (1)

**TFM** ↔ transfer-function matrix

**then-part** → consequence

**theoretical model** A mathematical model obtained with first-principles modelling, e.g., using conservation laws from the modelling domain. **S**: white-box model

**thermal actuator** An actuator, which produces a linear movement or a stroke caused by the expansion or contraction of heat-sensitive material. The latter is sealed in a housing with a diaphragm that pushes a piston within a guide. It can be implemented also as a part of MEMS. It is used in, e.g., temperature control, mixing, diverting, freeze protection, safety shutoff, valve actuation.

**thermal anemometer** → hot-wire anemometer

**thermal capacitance** → heat capacity

**thermal capacity** → heat capacity

**thermal conductance** The parameter of a mathematical model, describing the ability of an object e.g., a heat sink, to convey heat flow, defined as the ratio of heat flow to the temperature difference. It is the reciprocal of thermal resistance. **S**: absolute thermal conductivity

**thermal conductivity** The parameter of a mathematical model, describing the ability of homogenous material to convey heat flux, defined as the ratio of heat flux to the negative temperature gradient in the material. In the case of a two-dimensional barrier, the temperature gradient is the temperature difference on both sides of the barrier divided by the thickness of the barrier. It is the reciprocal of thermal resistivity. **S**: K-factor, K-value, specific thermal conductance

**thermal conductivity sensor** A sensor, which measures the heat flow through sample material according to the temperature difference between two points along the heat flow in the sample, considering the known geometry of the device, e.g., guarded hot-plate thermal conductivity sensor, comparative cut bar thermal conductivity sensor, hot-wire thermal conductivity sensor, MEMS thermal conductivity sensor. Fourier law enables the calculation of the heat-transfer coefficient.

**thermal flux** → heat flux

**thermal humidity sensor** A humidity sensor, which consists of two thermistors, one of which is hermetically sealed in a chamber with dry nitrogen and measures the dry-gas heat conductivity, while the other one is exposed to the measured gas passing through small holes measuring the heat conductivity of the gas containing water vapour. The difference of resistances of both thermistors is directly proportional to the measured absolute humidity. It is suitable for high temperature or high corrosive environments, e.g., drying kiln, oven, food dehydration plant, pharmaceutical plant.

**thermal imaging camera** A noncontact temperature sensor, which acquires a picture based on the infrared radiation from the observed object or scene. Its focusing lenses are made of special glass, which transmits infrared radiation. Infrared photodetectors measure the emitted radiation. It is used in, e.g., automotive night vision, noncontact temperature measurement, fire fighting, security. **S**: infrared camera, thermographic camera, TIC

**thermal insulance** The parameter of a mathematical model, describing the ability of a two-dimensional barrier to resist heat flux, defined as the ratio of the temperature difference on both sides of the barrier to heat flux. It is the reciprocal of thermal transmittance. **S**: R-factor, R-value

**thermal resistance** The parameter of a mathematical model, describing the ability of an object, e.g., a heat sink, to resist heat flow, defined as the ratio of the temperature difference to heat flow. It is the reciprocal of thermal conductance.

**thermal resistivity** The parameter of a mathematical model, describing the ability of homogenous material to resist heat flux, defined as the ratio of the negative temperature gradient to heat flux in the material. In the case of a two-dimensional barrier, the temperature gradient is the temperature difference on both sides of the barrier divided by the thickness of the barrier. It is the reciprocal of thermal conductivity. **S**: specific thermal resistance

**thermal switch** → temperature switch

**thermal transmittance** The parameter of a mathematical model, describing the ability of a two-dimensional barrier to convey heat flux, defined as the ratio of heat flux to the temperature difference on both sides of the barrier. It is the reciprocal of thermal insulance. **S**: U-factor, U-value

**thermistor** A temperature sensor, which is made of a temperature-sensitive semiconductor and exhibits a significant, precise and predictable change of resistance proportional to the temperature change. A mixture of metals, metal oxides, binders and stabilisers are pressed into a wafer and cut to the size of a chip. It is used in, e.g., temperature compensation, automotive applications, household appliances.

**thermocouple** A temperature sensor, which consists of two wires of dissimilar metals joined at one end that is placed where the temperature is measured, while the other end is kept on a constant temperature. A temperature-dependent voltage is generated in the circuit and converted to a useful signal. Numerous combinations of alloys allow selecting the best pair for a certain application. It enables robust measurements over a wide temperature range in many industrial, scientific and everyday appliances.

**thermographic camera** → thermal imaging camera

**thermometer** → temperature sensor

**thermopile** A temperature sensor, which converts thermal energy to electrical energy. Micromechanics and thin-film technology enable the design and manufacture of a miniaturised serially-interconnected array of thermocouples on a silicon chip. The absorbed radiated energy generates the output voltage proportional to the local temperature difference or temperature gradient. It is often used in noncontact temperature measurement applications.

**thermostat** A feedback control device, which measures the temperature of a physical system and maintains its temperature near the desired temperature, triggering corresponding heating or cooling actions. It contains various kinds of sensors, e.g., bimetal, thermistor, thermocouple, semiconductor, and is often used in, e.g., heating system, ventilation system, airconditioning system, as well as in water heater, thermostatic radiator valve, medical incubator, scientific incubator.

**thickness sensor** A sensor, which measures how thick is an immovable or a movable flat object, e.g., leather, paper, sheet metal, foil, plastic film, thin film. Both-sided measurements are conducted using two displacement sensors that are fixed opposite to each other on a frame. One-sided measurements are conducted without physical contact using, e.g., laser sensor, ultrasonic sensor, optical distance sensor. It is commonly used in, e.g., automation, control, QA.

**three-level control** The implementation of a control algorithm, the result of which is a control variable with a value from a set of three discrete values, which are usually 0, 1, and $-1$. **S**: three-step control

**three-level controller** A controller, the output of which has one of the three possible values, which are usually 0, 1, and −1. **S**: three-step controller

**three-mode controller** → proportional-integral-differential controller

**three-port valve** → three-way valve

**three-step control** → three-level control

**three-step controller** → three-level controller

**three-term controller** → proportional-integral-differential controller

**three-way valve** A valve, which is connected to three pipes. It can either have two inlets and one outlet or one inlet and two outlets. It enables fluid-flow mixing or fluid-flow splitting and usually uses an electric actuator, a pneumatic actuator or a thermal actuator. **S**: three-port valve

**threshold** **1**. The largest amount of measurement change, which produces no detectable reaction in the output of the measuring system. **2**. The value of a quantity, above or below which a certain event occurs.

**through-beam proximity sensor** A photoelectric proximity sensor or ultrasonic proximity sensor, which consists of an emitter that produces light beam or sound pulses, and a light receiver or a sound receiver. Both devices are installed in two opposing locations. An object, which passes between the two, breaks the light beam or sound pulses causing the reaction of the sensor output.

**through variable** A variable denoting a quantity, which is transmitted by an element, e.g., current, force, moment of inertia, flow, heat flow. It is measurable using the serial connection of the measuring instrument and the corresponding component, e.g., mass, rotational inertia, electrical coil.

**thruster** A propulsive device for controlling the pose, station keeping or long-duration acceleration of a spacecraft, e.g., rocket engine, ion thruster, plasma thruster or for improving manoeuvrability of a watercraft, e.g., azimuth thruster, bow thruster, stern thruster.

**thyristor** A multi-layer semiconductor bistable switch, which can also be used as an electric valve to control the electrical power. It allows current flow in only one direction and can be used as an element of, e.g., rectifier circuit, electric-motor speed control, heater control, power supply, actuator system. **S**: silicon-controlled rectifier

**TIC** ↔ thermal imaging camera

**tilt sensor** → inclinometer

**time constant** A parameter, which characterises the response of a first-order linear time-invariant system to a step input. It is the reciprocal value of the pole of the system. Its value is positive and has units of time, denoting the interval that is obtained either by measuring the time required for the step response to reach 63.2 % of the final value change, or by calculating the quotient of the final value change and the slope of the tangent of the response at the moment the input step occurs.

**time-constant transfer-function form** → Bode transfer-function form

**time delay** → dead time (1)

**time domain** Analytic space, in which the set of values that are accepted as the function input is covered by time as the independent variable.

**time-domain performance specifications** → transient-response specifications

**time-independent logic** → combinational logic

**time-invariant model** A mathematical model, the response of which depends on the shape of the input signal, but it does not depend on the start time of excitation, e.g., a model described by differential equations with constant coefficients.

**time response** The output-variable trajectory of a dynamic-system that is a consequence of the initial state or of the change in one of the input signals. The change may be caused by a standard test signal, e.g., step signal, ramp signal, pulse signal, sinusoidal signal, or a signal that is used in the regular operation of the system. The response is either measured, obtained by simulation, or calculated analytically. It is often represented graphically as a time-dependent function.

**time-response fitting** A method for the system discretisation, which results in a discrete-time model that has the same response to the chosen input signal in the sampling instants as the original continuous-time model, e.g., step invariance.

**time-rises** → rise time (1)

**time scaling** An algebraic operation, which converts the problem time defined in a mathematical model into the corresponding computer time by multiplying the first one by a dimensionless timescale factor. If the latter is greater than 1, the simulation is slower, while for the timescale factor smaller than 1, the simulation is faster with regard to the problem time. The procedure is required in analogue simulation but can be helpful also in digital simulation.

**time step** Difference between two consequent time samples.

**time-variant system** → time-varying system

**time-varying model** A mathematical model the response of which depends on the shape of the input signal as well as on the start time of excitation, e.g., model described by differential equations with time-varying coefficients.

220                                                                    21   T

**time-varying system** A system, the response of which depends not only on the shape of the input signal but also on the time of excitation occurrence, causing different responses to the same input signal at different time instants. Quantities governing its behaviour change with time due to, e.g., ageing, failures, environmental influences, structural changes. **S**: time-variant system

**TITO system**  → two-input two-output system

**torque balance** The momentum balance described with the modified second Newton's law. The time derivative of momentum expressed as the product of moment of inertia and angular acceleration is equal to the sum of external torques.

**torque—current analogy** An analogy between a rotational mechanical system and an electrical system, which results in additional analogue pairs, i.e., moment of inertia—capacitance, rotary damper—reciprocal of resistance, torsion spring—reciprocal of inductance, angular displacement—magnetic flux, angular velocity—voltage.

**torque sensor** A sensor, which converts torsional mechanical input into an electrical output signal, e. g., reaction torque sensor, rotary torque sensor. It is used for testing various devices, e.g., electric motor, fan, pump, automotive brake.

**torque—voltage analogy** An analogy between a rotational mechanical system and an electrical system, which results in additional analogue pairs, i.e., moment of inertia—inductance, rotary damper—resistance, torsion spring—reciprocal of capacitance, angular displacement—charge, angular velocity—current.

**torsional spring**  → torsion spring

**torsion spring** An idealised lumped-parameter element for modelling a rotational mechanical system, which stores energy. It does not introduce damping in the rotational mechanical system. It has the property of flexibility, e.g., torsion bar, torsion fibre, helical torsion spring. **S**: rotational spring, torsional spring

**touch sensor** A tactile sensor, which detects or measures the contact force at a defined point. It often provides binary-signal output.

**track drive robot**  → tracked robot

**tracked robot** A mobile robot, which uses metal or rubber tracks to move on rough, uneven, muddy or slippery surfaces. It provides excellent stability, flexible structure and low terrain pressure. **S**: track drive robot

**tracking control** Operation of a control system, which is primarily designed for the controlled variable to follow a changing setpoint. On the other hand, the disturbances are more or less neglected. **S**: setpoint control, variable-command control

**trajectory** 1. The curve, which describes the path of a point in space as a function of time. **2**. The curve, which describes the path of a point in the phase plane. **3**. The curve, which describes the path of a robot-manipulator end-effector or the path of a mobile robot.

**trajectory planning** → path planning

**transducer** A device, which transforms a physical quantity to another physical quantity, e.g., one form of energy to another form of energy. For instance, the measured temperature is transformed to pressure by a sensing element, the first converter transforms the pressure to displacement and the second converter transforms the displacement to the corresponding electrical signal, which is proportional to the measured temperature.

**transfer function** An input-output model of a linear time-invariant SISO system, expressed as the ratio of the corresponding transform, e.g., Laplace transform, Fourier transform, z-transform, of the output signal of the system to the corresponding transform of the input signal of the system, assuming zero initial conditions. It can appear in the polynomial transfer-function form, in the zero-pole-gain transfer-function form or in the Bode transfer-function form.

**transfer-function matrix** An input-output model of a linear time-invariant MIMO system described with the matrix of rational polynomials, which is a generalisation of the SISO transfer function. Its elements relate each input of actuators to each input of sensors. Consequently, cross couplings are the off-diagonal elements, while the transfer functions of direct input-output pairs are the diagonal elements. It has as many rows as outputs and as many columns as inputs. **S**: TFM

**transfer-function parallel decomposition** A method, which reformulates a complex transfer function to a sum of simpler transfer functions, often of the first order for real poles and of the second order for complex-conjugate pole pairs. It is frequently used for the development of a simulation scheme for a mathematical model described by a transfer function.

**transfer-function serial decomposition** A method, which reformulates a complex transfer function to a product of simpler transfer functions, often of the first order for real poles and of the second order for complex-conjugate pole pairs. It is frequently used for the development of a simulation scheme for a mathematical model described by a transfer function.

**transform** 1. A relation of a function in one domain to another function in an alternative domain, e.g., Laplace transform. **2**. The result of a mathematical operation that relates two domains, e.g., Laplace transform of a time-domain variable.

**transient contact** A contact between an operator and a part of a robot system, in which the robot hits the operator with a short impact. The body part of the operator is not clamped and can recoil or retract from the moving part of the robot system.

**transient response** 1. The response of a dynamic system to any change of a steady state or to any change of an equilibrium position. **2**. The response of a dynamic system, which is present in a short time period after the system is turned on. For an asymptotically stable system, it disappears, while for an unstable system, it increases indefinitely. **3**. The part of a dynamic-system response to any change of input, which lasts until the system reaches a new steady state.

**transient-response specifications** Indicators, which define the performance characteristics of a control system according to the transient response of a proportional, underdamped second-order system to a unit-step signal, i.e., delay time, rise time, peak time, maximum overshoot, reaching time and settling time. They are defined for a second-order system because they are analytically calculable, while for the higher-order system they are not in common use. **S**: time-domain performance specifications

**transistor power amplifier** An electrical multistage amplifier, the output current of which is much greater than the input current, while its output voltage is usually lower than the input voltage. It is often in a form of an integrated circuit and is used in, e.g., electric-motor drives.

**transit-time ultrasonic flow meter** An ultrasonic flow meter, which bases on the difference between the upstream and the downstream ultrasound propagation velocities. It utilises two transducers incorporating both a transmitter and a receiver. Alternating the transmitting and the receiving of ultrasonic waves between the two transducers results in a time difference, which is proportional to the measured fluid velocity and consequently to the flow rate. It can be used for any homogeneous fluid that is capable of propagating ultrasonic waves.

**translational damper** An idealised lumped-parameter element for modelling a translational mechanical system, which dissipates kinetic energy as the consequence of energy conversion. It decreases the oscillations in a translational mechanical system and has the property of resistance, e.g., shock absorber, dashpot.

**translational joint** $\rightarrow$ prismatic joint

**translational mass** An idealised lumped-parameter element for modelling a translational mechanical system. It stores kinetic or potential energy and has the property of inertia. It is essential to any translational mechanical system.

**translational mechanical system** A system, which consists of idealised, weightless and dimensionless elements for linear motion, e.g., mass, spring, damper.

**translational motion** $\rightarrow$ linear motion

**translational spring** An idealised lumped-parameter element for modelling a translational mechanical system, which stores potential energy. It does not introduce damping in a translational mechanical system and has the property of flexibility.

**transmission zero** The zero of a transfer function in the diagonal of the Smith-McMillan canonical form of a TFM of a linear MIMO system. No pole-zero cancellations occur in the TFM, even though some zeros of some TFM elements are equal to some poles of some other TFM elements. **S**: blocking zero

**transmitter** 1. A device, which converts the output signal of a sensor into an analogue or digital standardised instrumentation signal. It is often integrated with a sensor into a single device. **2.** A telecommunication device for emitting electromagnetic signals, often used in spatially dispersed systems for conveying signals.

**transportation lag** $\rightarrow$ dead time (1)

**transport delay** $\rightarrow$ dead time (1)

**transpose Jacobian control** Control of a robot, which relates the forces at the robot end-effector with the robot joint torques. When the robot moves away from the desired pose, the springs pull the robot end-effector into the desired pose with a force proportional to the positional error.

**triac** An electrical element, which consists of two thyristors, connected in inverse parallel to each other, allowing current in both directions. It is widely used in AC-power control applications, e.g., lighting control, fan control, small electric motor control as well as in actuator system.

**trial and error** A fundamental, nonsystematic problem-solving technique, which is used to find a solution by repeating varied attempts until success or until the procedure is stopped in some other way. It is successfully implemented in simple problems or when no apparent rule applies and is often used by users who have little knowledge in the problem domain. **S**: cut-and-try

**truncation error** An error of the numerical integration method, which occurs due to inherent limitations of a certain integration algorithm. It is independent of the numerical accuracy of the used digital computer. **S**: numerical approximation error

**tuning-fork level switch** A level switch that consists of a piezoelectric oscillator, which vibrates a metal resonator with two prongs at its natural frequency for air, and of a piezoelectric detector that senses the change of frequency when the detected material covers the resonator. It is used for point level detection of liquids and bulk materials in applications dealing with, e.g., sticky, corrosive, perturbed, splashing, foaming, viscous liquids as well as with powders and fine-grained bulk solids. **S**: vibrating-fork level switch

**tuning rules** Rules, which enable the calculation of mostly PID-type-controller parameters, based on estimated or measured data about the dynamics of the controlled system. Either a mathematical model of the system can be used or the tuning is performed by experimenting with the real system. They are frequently provided as a table and are often named after their authors, e.g., Ziegler-Nichols rules, Chien-Hrones-Reswick rules, Cohen-Coon rules, Opelt rules, Åström-Häglund rules.

**turbidimeter** A photometer, which measures the loss of transmitted light intensity due to the scattering effect of particles suspended in a solution. A light of known wavelength is passed through a cuvette containing the sample. A photomultiplier tube measures the transmitted light, which is a function of the concentration of the suspended substance. It is used in, e.g., air pollution measurement, water pollution measurement, chemical industry, pharmaceutical industry, cosmetic industry, food industry, medicine. **S**: opacimeter

**turbine flow meter** A flow meter, which measures volumetric flow rate. It consists of a free-spinning rotor, driven by the measured flow rate. When magnetic blades pass an external pickup mounted on the body of the sensor, a frequency signal is generated. The frequency is proportional to the angular velocity of the rotor and to the velocity of the flow and consequently to the measured flow rate. It also straightens the flow, thus minimising the negative effects of turbulence. It is often used for measuring, e.g., petrochemicals, organic liquids, inorganic liquids, liquified gases.

**Tustin's method** $\rightarrow$ bilinear transformation

**twist** A screw, representing motion, expressed by a pair of three-dimensional linear velocity and angular velocity vectors, which point along or around the line of action.

**two-input two-output system** A MIMO system with two inputs and two outputs. It is often used as an illustrative example for MIMO systems. **S**: TITO system

**two-level control** $\rightarrow$ on-off control

**two-level controller** $\rightarrow$ on-off controller

**two-seat valve** $\rightarrow$ double-seat valve (1)

**two-step control** $\rightarrow$ on-off control

**two-step controller** $\rightarrow$ on-off controller

**type number** $\rightarrow$ system type

# Chapter 22
# U

**UAV** ↔ unmanned air-vehicle

**uC** → microcontroller

**U-factor** → thermal transmittance

**ultimate gain** The gain value that causes the output of a control loop to achieve sustained constant oscillations. It is often used in Ziegler-Nichols closed-loop tuning method. **S**: critical gain

**ultimate-gain method** → Ziegler-Nichols closed-loop tuning method

**ultimate period** A period of sustained oscillation for an undamped, marginally stable system. It is often used in Ziegler-Nichols closed-loop tuning method. **S**: critical period

**ultrasonic density meter** A density meter, which measures the consistency of materials, solid-suspended liquids, sludges or slurries. On one hand, it can transmit sound waves through a sample and measure the attenuated and the dispersed waves by a receiver. On the other hand, it can calculate the sound velocity, which is proportional to the measured density, by transmitting sound waves through the sample to a receiver and measuring the time-of-flight, considering the known distance between the transmitter and the receiver.

**ultrasonic distance sensor** A distance sensor, which uses sound waves time-of-flight principle to determine the distance to the measured object. It is used in, e.g., chemical industries, petroleum industries, manufacturing industries as well as in devices, e.g., robot, smart car, UAV.

**ultrasonic flow meter** An obstructionless flow meter, which measures the velocity and consequently the flow rate of clear liquids, dirty aerated liquids, slurries and gases by detecting the transmitted or reflected acoustic vibrations, e.g., transit-time

© ZRC SAZU/Research Centre of the Slovenian Academy of Sciences and Arts 2023
R. Karba et al., *Terminological Dictionary of Automatic Control, Systems and Robotics*,
Intelligent Systems, Control and Automation: Science and Engineering 104,
https://doi.org/10.1007/978-3-031-35755-8_22

ultrasonic flow meter, Doppler ultrasonic flow meter. It can be mounted inside the flowstream. Alternatively, it can be either portable and clamped on the pipe, or permanently fixed on the outside of the pipe, without having contact with the measured fluid.

**ultrasonic level sensor**  A level sensor, which measures the time-of-flight of sound waves. A one-piece oscillator emits and receives waves alternately. The waves reflect like an echo from the surface of the measured fluid, i.e., from the border between two media with sufficiently different densities. The time elapsed between the pulse emission and reception is proportional to the measured fluid level in a container with known geometry. It enables noncontact continuous measurements and can be used also in applications dealing with aggressive, corrosive or abrasive media in harsh environmental conditions. S: ultrasonic level transmitter

**ultrasonic level transmitter**  → ultrasonic level sensor

**ultrasonic motor**  An electric motor, which exploits the friction between a vibrating stator and a rotor or a slider. It can provide either rotational or translational movements. The resonant high-frequency vibrations are caused by a piezoelectric crystal. Despite its compact construction, it provides relatively high torque, as well as silent and accurate operation. It is often used in cameras to assure smooth, quiet and fast focusing. S: silent wave motor

**ultrasonic proximity sensor**  A proximity sensor, which uses the transmitted or reflected sound waves to detect the presence of a target object, e.g., through-beam ultrasonic proximity sensor, retro-reflective proximity sensor, diffuse ultrasonic proximity sensor. It can also detect irregularly shaped objects made of various materials in a variety of operating conditions.

**ultrasonic sensor**  A sensor, which measures the distance to surrounding objects using propagation of sound above 20 kHz. It emits a sound wave and receives its reflection from the object. The distance is calculated from the time difference between the emission and the reception. S: sonar (2)

**uncertainty**  The difference between the considered system behaviour and its model behaviour. The distinctions can be the consequence of incomplete knowledge of the system, of some simplifications in the model or of discrepancies in the structure of the system and its model.

**unconstrained optimisation method**  An optimisation method, which seeks the global minimum, the global maximum, a local minimum or a local maximum, in the absence of restrictions, e.g., gradient method, direct search method, simplex method. Its optimised parameters are not limited.

**uncrewed aerial vehicle**  → unmanned air-vehicle

**undamped frequency**  → natural frequency

## 22 U

**undamped oscillation** Oscillatory time response, which has a constant amplitude. It has no energy loss or it is correspondingly compensated. Its damping factor is therefore equal to 0.

**underactuated manipulator** A robot, which has a lower number of actuators and fewer input commands than DOF, with at least one passive DOF. If control is feasible according to the prescribed task, its weight, cost and energy consumption can be reduced. Its flexible structure and performance allows its application in, e.g., medical robot, unmanned system. **S**: underactuated robot

**underactuated robot** → underactuated manipulator

**under-damping** Damping, where the step response of a system reaches the new steady state by oscillating with the corresponding decay rate. In such a case the damping factor is smaller than 1.

**underwater mobile system** A mobile system, which is submerged below a liquid surface, e,g., submarine, autonomous underwater vehicle. It can operate either with or without a pilot.

**unforced response** → natural response

**uniform white noise** A white noise, which has the probability density function equal to that of an uniform distribution. In practice, the generated noise is white noise only within the frequency range of interest.

**uninterruptible power supply** A backup power supply, which in the case of power failure or big power fluctuations comes on, enabling undisturbed operation of a system till the main power supply starts working again. It consists of an energy-storing element and a corresponding power converter, e.g., rechargeable battery, inverter. It is used to protect hardware, e.g., computer, data centre, control equipment, telecommunication equipment, against unexpected power disruption that can cause serious damage. **S**: UPS

**unitary PLC** → compact PLC

**unit factor method** → dimensional analysis (1, 2)

**unit impulse** A brief test signal described by the Dirac delta function.

**unit impulse function** → Dirac delta function

**unit ramp function** A function, which enables modelling evenly increasing test signal in dynamic system analyses. It has 0 value at negative values of the independent variable and a constant slope of 1 at positive values of the independent variable.

**unit ramp signal** An aperiodic standard signal for testing the behaviour of dynamic systems, which is modelled by a unit ramp function.

**unit step function**  A function, which enables modelling of a quickly and abruptly changing test signal in dynamic-system analyses. Its value is 0 at negative values of the independent variable, and 1 at positive values of the independent variable. **S**: Heaviside function

**unit step signal**  An aperiodic standard signal for testing the behaviour of dynamic systems, which is modelled by a unit step function.

**unity feedback loop**  A feedback loop, which has no signal modifiers in the feedback path. Hence, the gain of the feedback path is 1.

**unity-gain frequency**  $\rightarrow$ gain crossover frequency

**univariable system**  $\rightarrow$ single-input single-output system

**universal approximator**  The mathematical model composed of basis functions that can approximate any continuous function to the specified accuracy provided that the finite number of basis functions is large enough, e.g., ANN, fuzzy model.

**universal joint**  **1**. A mechanical link between two intersecting nonparallel shafts. **2**. A robot joint, which consists of two perpendicular rotations. It can be found as a passive joint in a parallel robot.

**unmanned air-vehicle**  A self-powered aircraft or missile, which is controlled by an onboard computer or is controlled by a ground pilot or a pilot in another aircraft, e.g., remotely controlled aeroplane, remotely controlled helicopter. **S**: drone (2), remotely-piloted aircraft, UAV, uncrewed aerial vehicle

**unscented Kalman filter**  An algorithm for the optimal estimation of noisy nonlinear dynamic-system states. For variables with non-Gaussian distributions, the algorithm uses the approximate distributions of the estimates of the variables.

**unscented transform**  A mathematical function for estimating the result of the operation, which applies a given nonlinear transformation to a probability distribution. The latter is defined with statistical parameters, which are derived from a finite dataset. **S**: unscented transformation

**unscented transformation**  $\rightarrow$ unscented transform

**unstructured environment**  Robot surroundings, the geometrical or physical characteristics of which are not known apriori.

**unstructured uncertainty**  A parametric uncertainty, which is visible as a difference between the system behaviour and its model behaviour. The exact location of the uncertainty in the structure is not known.

**unsupervised learning**  A family of machine learning methods used for determining the function from a set of functions, which best describes the data that is not apriori segmented to input data and output data. It is used in, e.g., data compression, clustering.

**UPS** ↔ uninterruptible power supply

**utility function** → fitness function

**U-tube manometer** A pressure sensor, which consists of two vertical glass or acrylic tubes connected at the bottom. It contains water, oil or mercury. It can measure absolute pressure, differential pressure or vacuum. The difference between the levels of liquid in the tubes is proportional to the measured quantity. The inclination of the device can improve its resolution in the case of a small pressure difference.

**U-value** → thermal transmittance

# Chapter 23
# V

**vacuum pump** A pump, which removes gas molecules from a sealed chamber to achieve extremely low gas pressure, e.g., rotary vane pump, radial piston pump, diffusion pump, turbomolecular pump, ion pump. It is used in, e.g., vacuum engineering, photolithography, mass spectroscopy, electronic microscopy

**validity function** → membership function

**validity of concepts** A model-validation procedure, which, on one hand, tests the basic assumptions in the process of model development, and, on the other hand, collects the data supporting the applied natural laws.

**validity of data** A model-validation procedure, which tests the quality of data that can be questionable due to different reasons, e.g., observation errors, calibration errors, interpolation or extrapolation, inaccurate parameter estimation.

**validity of inference** A model-validation procedure, where a number of relevant participants in the modelling process, e.g., modellers, domain experts, users, consent that the model results provide the same conclusions.

**validity of methodology** A model-validation procedure, which tests the appropriateness of the applied techniques, e.g., justification of the linear approximation of a system, justification of a discrete-equivalent representation of a continuous system, suitability of numerical or computational methods.

**validity of results** → empirical validity

**valve** A device, which controls the flow of fluid or energy by directly affecting the controlled object. It can be either of electrical type, e.g., transistor, thyristor, rheostat, or process type, e.g., control valve, on-off valve, safety valve.

**valve actuator** An actuator, which uses a power source to enable valve-stem positioning, e.g., electric actuator, pneumatic actuator, hydraulic actuator, manual actuator. It must be adapted to the signal of the controller as well as to the valve-stem,

© ZRC SAZU/Research Centre of the Slovenian Academy of Sciences and Arts 2023
R. Karba et al., *Terminological Dictionary of Automatic Control, Systems and Robotics*,
Intelligent Systems, Control and Automation: Science and Engineering 104,
https://doi.org/10.1007/978-3-031-35755-8_23

e.g., linear or rotary motion, on-off valve or control valve, globe valve or diaphragm valve. Application-specific requirements must also be taken into account, e.g., speed of actuation, need for manual override, environmental factors.

**valve body** A part of a valve, which holds together all the other parts, enabling the connection to the piping. It is usually cast or forged from metal, e.g., cast iron, stainless steel, alloy, bronze, brass, titanium, or is made of plastic, e.g., polyvinyl chloride, polyvinylidene fluoride, polypropylene, fibreglass.

**valve disc** A movable obstruction in the valve body, which adjustably restricts the flow through the valve, moving linearly or rotating on a valve stem. It is often forged and hard-surfaced to achieve the required properties. **S**: valve member

**valve flashing** A thermodynamic process, which occurs when the flow of liquid encounters an obstacle or contraction. This may cause an increase of flow velocity and consequently a decrease of pressure below the vapour pressure. Therefore, evaporation begins and vapour bubbles are formed. When the bubbles impinge on valve components, they may cause corrosive damage even when no abrasive solids are present in the flow.

**valve lift** → valve travel (1, 2)

**valve member** → valve disc

**valve packing** The seal between the valve stem and the valve body, which prevents leakage of the process fluid. It is made of, e.g., elastomers, polyacrylate, silicone, fibrous materials. It must allow the motion of the valve stem, caused by some external device, e.g., actuator, handwheel. **S**: valve stem seal

**valve port** A passage, which allows the fluid to pass through the valve, according to the position of the valve disc. It enables the connection of the valve body to one pipe, to several pipes or some other components.

**valve positioner** A slave-controller, which compares the valve-stem position-feedback signal and the setpoint, e.g., pneumatic valve positioner, electric valve positioner, analogue electro-pneumatic valve positioner, digital valve positioner. It is connected mechanically to the valve stem enabling precise operation of the valve and consequently the optimal performance of the controlled plant. It is implemented in, e.g., automotive industry, aerospace industry, marine industry, military industry, food industry, pharmaceutical industry.

**valve rangeability** The ratio between the maximum and the minimum flow through a valve, which is important in the process of valve-type selection. Often, a high value is required, which assures that the valve correspondingly controls small as well as large flows. It defines the initial valve throat area for the valves with equal-percentage characteristic.

**valve seat** The single or multiple interior part of a valve body, which contacts one valve disc or several valve discs, e.g. hard valve seat, soft valve seat.

**valve sizing coefficient** → flow coefficient

**valve spindle** → valve stem

**valve stem** The connection between the valve disc and the handwheel or lever, in manual control, or between the valve disc and the actuator, in automatic control. The transmitted motion may be caused by a force or by a torque. **S**: valve spindle

**valve stem seal** → valve packing

**valve stroke** → valve travel (1, 2)

**valve throat area** The area between the valve disc and the valve seat, which lets the fluid to flow through. Its relationship to the flow rate is always directly proportional. However, different valve travels can result in different flow rates, depending on the shape of the valve disc. **S**: orifice pass area

**valve travel** 1. The distance passed by the valve disc when it moves from completely closed to the completely opened position. It can be expressed by linear displacement or angular displacement of the valve disc. **S**: valve lift, valve stroke 2. The relative position of the valve disc from its closed position, often given in percents. **S**: valve lift, valve stroke

**0-10 V analogue signal** A standardised electrical analogue signal transmission, which is susceptible to electrical interference and voltage drops, caused by wire resistance.

**variable-area flowmeter** → rotameter

**variable-command control** → tracking control

**variable-frequency drive** → variable-speed drive

**variable reluctance sensor** A sensor, which measures linear displacement or linear velocity, angular displacement or angular velocity, or proximity of a ferrous object. The measurement is based on changing magnetic reluctance, e.g., a ferromagnetic cogwheel on a rotating shaft.

**variable-speed drive** An electrical device, which controls AC-motor speed and torque by varying the motor's input frequency and voltage, enabling its smooth startup. Motor voltage and frequency are often varied by PWM. It appears in, e.g., control, heating, ventilation, air conditioning. **S**: adjustable-frequency drive, inverter drive, variable-frequency drive, variable-voltage variable-frequency drive, VSD

**variable-step integration method** A numerical integration method, which enables automatic adjustment of step size during the simulation run according to the dynamics of the simulated process and the required tolerance.

**variable stiffness actuator** An actuator, in which the mechanical impedance can be adjusted by changing the parameters of a robot joint. It can be used to make the robot safer in the case of collision as the robot joint stiffness and impact of inertia are reduced. **S**: VSA

**variable-structure control** A nonlinear control system consisting of a set of continuously operating subsystems and specific rules, which govern the switching from one subsystem to another, causing discontinuities in the control signal. It substantially broadens control capabilities enabling various practical applications in, e.g., mobile robot, electric drive, spacecraft, underwater vehicle.

**variable-voltage variable-frequency drive** → variable-speed drive

**24 V digital signal** A standardised digital signal transmission with low energy consumption, which is directly compatible with semiconductor technology and can use a common power supply.

**vector optimisation** → multi-objective optimisation

**velocity-error constant** **1**. A closed-loop system parameter, which is defined by the limit value of the open-loop transfer function multiplied by $s$ as $s$ approaches 0. Here, $s$ is the complex variable in the $s$-plane. **2**. A closed-loop system parameter, the value of which is inversely proportional to the steady-state error of the response of the system to a unit-ramp signal.

**vena contracta** A spot in a flowstream where its cross-section is minimal, the fluid velocity is maximal and, consequently, the pressure is minimal. It takes place slightly downstream of the orifice or other restriction with sharp edges, which causes this phenomenon. It is crucial in the design of a differential-pressure flow meter influencing the spots where the pressure difference is measured.

**Venturi flow meter** A differential-pressure flow meter, which consists of three sections of a pipe. A construction, assembled from a converging cone, a throat and a diverging cone, enables a large flow rate with a small pressure drop. It is also suitable for liquids that contain a large share of solid particles.

**verbal model** A symbolic model, which gives spoken or written description of the studied system or its behaviour. It is often difficult to obtain unambigous and accurate information from such a model. **S**: linguistic model

**verification** A procedure, which evaluates the design consistency of a mathematical model including the accuracy and correctness of modelling methodologies, algorithms, computer programs etc. At this stage, the model behaviour is not directly compared to the real-system behaviour.

**vibrating-fork level switch** → tuning-fork level switch

**vibrating-rod level switch** A level switch, which consists of a metal pole excited by a piezoelectric oscillator to resonate at its natural frequency for air. When the measured material touches the pole, the amplitude of vibrations falls below the threshold and the output signal changes. It is used for point level detection of liquids and bulk materials in applications dealing with, e.g., powder, cement, salt, sand, flour, dry cereals, tobacco, plastic granules, wood chips. **S**: vibronic point level detector

**vibrating-tube density meter** → oscillating U-tube

**vibrating-wire piezometer** An absolute pressure sensor, which consists of a fine high tensile-strength steel wire fixed at one end and attached to a diaphragm in contact with the measured pressure at the other end. Excited with a piezoelectric actuator, it oscillates at a frequency proportional to the tension of the wire enabling the determination of the pressure. It is immune to electrical noise and needs temperature compensation.

**vibrational viscometer** A viscometer, which measures the viscosity of Newtonian fluids as well as non-Newtonian fluids. It consists of a stainless steel rod submerged in a tested fluid that oscillates at its natural frequency. The measured viscosity can be determined either from the power input necessary to keep a constant amplitude of oscillations, from the time, in which the oscillations degrade once the excitation is switched off or from the frequency, at which a corresponding phase shift between the excitation waves and the response waves is achieved. It is suitable for various fluids, e.g., bio-fuel, oil, beverage, coating, polymer, wastewater.

**vibration gyroscope** A gyroscope, which measures angular velocity from the Coriolis force applied to an oscillating object. It is often implemented as MEMS and used in, e.g., smartphone, digital camera, robot, car navigation, sports sensor, mobile game, as well as for motion sensing.

**vibronic point level detector** → vibrating-rod level switch

**violet noise** Coloured noise, the power spectral density of which is directly proportional to the square of the frequency. Therefore, its power spectral density increases by 20 dB/decade. **S**: differentiated white noise, purple noise

**virtual coupling** A virtual mechanical system implemented as a mathematical model of a spring-damper structure, which is interposed as a layer between a haptic interface and a virtual environment to guarantee the stability of the system.

**virtual environment** An interactive, computer-generated model of the real environment. It can be constructed as an authentic model of the real environment or can be a highly simplified version of reality.

**virtual-environment rendering** A rendering, which generates visual, audio and haptic information presented to the user through a corresponding display, a loudspeaker or a haptic interface.

**virtual fixture** Force and position signals generated by the software for a human operator. The signals represent the constraints of robot movements. It improves safety, accuracy and speed of robot-assisted manipulation tasks and is frequently used in teleoperation systems and collaborative operations.

**virtual prototyping** A software-based technology, which enables modelling of a system, as well as simulation and 3D visualisation of its behaviour in realistic operating conditions. It is used iteratively and substitutes rapid prototyping in the design process. It uses CAD software and CAE software to optimise and validate the design before a possible physical prototype is built. Numerous design variants can easily be tested on a computer, which improves the human-product interaction. **S**: digital prototyping

**virtual sensor** → soft sensor

**viscometer** A sensor, which measures the resistance of a fluid to gradual deformation, e.g. rheometer, falling-ball viscometer, falling-piston viscometer, rotational viscometer, vibrational viscometer, capillary viscometer. The resistance is caused by internal friction forces between adjacent layers of a fluid flow in a tube at a given temperature. A force needed to maintain the flow in a tube is proportional to the measured dynamic viscosity or kinematic viscosity. It is used in various industries to monitor batch consistency and QC. **S**: viscosimeter

**viscosimeter** → viscometer

**vision-guided robot** A robot, which uses optical information obtained from a corresponding vision system, which provides feedback to the robot controller, enabling an accurate execution of the given task.

**vision system** A system, which comprises optics, illumination and sensors, e.g., digital camera, high-speed digital camera, laser sensor, as well as a microprocessor or a computer and associated software.

**visual modelling** The development of a schematic model, which graphically represents applications in a human-readable form, using block-oriented simulation languages with pre-built components, e.g., Modelica, Simulink. It enables effective communication among designers.

**visual servoing** A vision-based robot or mobile system control, which uses visual feedback information to achieve the desired motion. It can be position-based, image-based or hybrid. The corresponding sensor may be carried by the robot or it may be fixed in the environment.

**volume flow rate** → volumetric flow rate

**volumetric flow rate** The time derivative of the volume of the substance being transferred through a predefined surface or through an object, e.g., an orifice. **S**: rate of fluid flow, volume flow rate, volume velocity

**volume velocity** $\rightarrow$ volumetric flow rate

**vortex flow meter** A flow meter, which measures the volumetric flow rate in a pipe by using a square, rectangular, trapezoidal or T-shaped shedder bar for generating a series of vortices. The occurrence frequency of vortices is proportional to the flow velocity and consequently to the measured flow rate. The presence of vortices is detected using different sensors, e.g., piezoelectric sensor, ultrasonic sensor, capacitive sensor. It has a simple structure, no moving parts and is therefore very reliable. It is used for the measurement of steam flow, gas flow or liquid flow.

**VSA** $\leftrightarrow$ variable stiffness actuator

**VSD** $\leftrightarrow$ variable-speed drive

# Chapter 24
# W

**water mobile system** A mobile system, which floats on the liquid surface, e,g., ship, boat, sailing boat. It can operate either with or without a pilot.

**wearable robot** → robotic exoskeleton (1, 2)

**wearable sensor** A sensor, which is applied directly on skin or indirectly by integration in various accessories, e.g., fashion accessories, garments, shoes, wristwatches, smartphones. Physiological monitoring is enabled using, e.g., wearable mechanical sensor, wearable electrical sensor, wearable optical sensor, wearable chemical sensor.

**wedging** The state, in which the object rests in an undesired orientation because it is improperly grasped by the robot fingers. The object is therefore stuck due to its shape.

**weighing scales** A device, which measures the mass of an object by converting its gravitational force into an electric signal, e.g., balance scales, load cell, spring scales, electromagnetic scales, digital scales. It is commonly used in, e.g., QC, bulk material handling, high precision weighing, filling, dosing, dispensing.

**weight function** **1**. A mathematical function that defines the influence of an element on the final result, e.g., in objective function, in convolution integral. **S**: weighting function **2**. The impulse response of a dynamic system. **S**: weighting function

**weighting factor** A constant value of a weight function, pondering the importance of a certain element in a group.

**weighting function** → weight function (1, 2)

**wet-test gas flow meter** A gas flow meter, which consists of a drum sealed with liquid and a rotating impeller with specially shaped vanes that pass fixed volumes of gas towards the output. Most kinds of gas can be measured without the influence of composition and condition, e.g., density, viscosity. **S**: drum-type gas flow meter

**wheeled robot** A mobile robot, which moves on usually flat ground using a number of rotating circularly-shaped discs. The possible realisations depend on their number, type, implementation, geometric characteristics and motorisation.

**white-box model** $\rightarrow$ theoretical model

**white noise** A noise with a constant power spectral density. Its power is equal within any equally sized, frequency band.

**Whittaker-Shannon interpolation** A method for obtaining the ideal reconstruction of a frequency-limited continuous-time signal from its discrete-time samples.

**Wiener filter** A filter, which produces reduced-noise estimates of the observed noisy process behaviour by linear time-invariant filtering. It is given in the input-output form and is used for noise reduction in, e.g., control, system identification, signal processing, image processing.

**Wiener model** A block-structured nonlinear mathematical model, which is made up of two submodels connected in series, where the first one is a dynamic linear submodel, which is connected to the input of static nonlinear mapping. It is often combined with the Hammerstein model into various block structures.

**wireless sensor network** A connected set of spatially dispersed sensors for monitoring and recording physical or environmental conditions, using electromagnetic communication. It consists of nodes that include sensors, processing units, communication modules and power supplies. It is used in, e.g., process monitoring, control monitoring, pollution monitoring, wastewater monitoring, healthcare.

**wireless wearable sensor** A miniature wearable sensor, which uses radio communication and contactless power transfer, enabling continuous real-time sensing and integration in wireless sensor networks or in IIoT. It is used in, e.g., digital healthcare, rehabilitation, sports, military.

**working range** $\rightarrow$ measuring range

**world coordinate frame** A Cartesian coordinate frame, which has its origin in a fixed position, usually connected to a particular task. It is defined by the user, with $z$-axis pointing upwards. **S**: robot-task frame

**wrench** A screw, representing loading, expressed by a pair of three-dimensional force and torque vectors, which point along or around the line of action.

# Chapter 25
# Y

**yaw angle**  The angle, which determines the rotation around the vertical axis of an object or a mobile system, e.g., aeroplane, ship, robot end-effector. Besides the pitch angle and roll angle, it is an element from the set of three angles that completely determine the orientation of an object in space. **S**: heading

# Chapter 26
# Z

**zero** **1**. A value of complex variable $s$ or complex variable $z$, which results in a zero-valued numerator of the continuous or discrete transfer function. Its position in the complex plane affects the behaviour of the system. **2**. A root of the numerator of a transfer function. Its position in the complex plane affects the behaviour of the system.

**zero assignment** A controller-design method for a linear MIMO system. The parameters of the controller are calculated according to the requirements concerning the predefined locations of some or all the closed-loop-system zeros in the $s$-plane. As feedback controllers do not affect the zeros of the system, a corresponding static cascade controller must be applied.

**zero error** The reading of a measuring instrument at zero input when it should be 0. It is often the consequence of bad calibration.

**zero-moment point** A principle, which is used to generate stable walking patterns in legged locomotion, e.g., for a humanoid robot. It is the point where the resultant of ground reaction forces intersects with the ground. It is also the point on the ground surface where the net angular momentum is equal to 0. **S**: ZMP

**zero-order hold** An element that converts a discrete-time signal to a continuous-time signal by keeping its sampled value constant from the actual sampling instant until the next sampling instant. **S**: ZOH

**zero-pole-gain transfer-function form** The structure of a transfer function, which has a product of multiplicative constant and sums of complex variable $s$ or complex variable $z$ with the negative values of zeros in its numerator. The product of sums of complex variable $s$ or complex variable $z$ with the negative values of poles is in its denominator. **S**: factored transfer-function form

**zero-pole mapping** A method for the system discretisation, which results in a discrete-time model that has poles and zeros directly calculated from the poles and zeros of the original continuous-time model.

© ZRC SAZU/Research Centre of the Slovenian Academy of Sciences and Arts 2023
R. Karba et al., *Terminological Dictionary of Automatic Control, Systems and Robotics*,
Intelligent Systems, Control and Automation: Science and Engineering 104,
https://doi.org/10.1007/978-3-031-35755-8_26

**Ziegler-Nichols closed-loop tuning method** A method for tuning PID-type controllers in a closed loop. By increasing the gain of the open-loop-stable system with the dead time or with at least third-order dynamics, a sustained oscillation with constant amplitude is achieved. The parameters of the controller can be calculated from the estimated ultimate gain and the ultimate period using the corresponding table with the empirical formulas. **S**: ultimate-gain method, Ziegler-Nichols oscillation method, Ziegler-Nichols second method

**Ziegler-Nichols first method** $\rightarrow$ Ziegler-Nichols open-loop tuning method

**Ziegler-Nichols open-loop tuning method** A step-response method that enables the calculation of the parameters of a PID-type controller according to the estimated gain of the system, dead time and rise time using the corresponding table with the empirical formulas. **S**: Ziegler-Nichols first method

**Ziegler-Nichols oscillation method** $\rightarrow$ Ziegler-Nichols closed-loop tuning method

**Ziegler-Nichols rule** A heuristic tuning rule for PID-type controllers that comprises the Ziegler-Nichols closed-loop tuning method and the Zigler-Nichols open-loop tuning method.

**Ziegler-Nichols second method** $\rightarrow$ Ziegler-Nichols closed-loop tuning method

**ZMP** $\leftrightarrow$ zero-moment point

**ZOH** $\leftrightarrow$ zero-order hold

**ZOH equivalent** $\rightarrow$ step invariance

**$z$-plane** The analytic space of complex variable $z$, which is used for the presentation of, e.g., poles and zeros of a transfer function of a discrete-time system. Its $x$-axis depicts the real parts and its $y$-axis depicts the imaginary parts of the complex numbers $z$. It is a graphical analysis tool in engineering and physics.

# Part III
# References

# Chapter 27
# References and Terminological References

## References

1. Albertos Pérez P, Antonio S (2004) Multivariable control systems: an engineering approach. Springer
2. Angelov P, Filev DP, Kasabov N (eds) (2010) Evolving intelligent systems: methodology and applications. Wiley
3. Åström KJ, Wittenmark B (2008) Adaptive control. Mineola
4. Atherton D, Borne P (eds) (1991) Concise encyclopedia of modelling and simulation. AdvSyst Control Inf Eng
5. Baillieul J, Samad T (eds) (2021) Encyclopedia of systems and control. Springer
6. Battikha NE (1997) Condensed handbook of measurement and control. Instrumentation System and Automation Society
7. Bajd T, Mihelj M, Munih M (2013) Robotics. Springer
8. Bajd T, Mihelj M, Lenarčič J, Stanovnik A, Munih M (2010) Robotics. Springer Science & Business Media
9. Borer J (1991) Microprocessors in process control. Elsevier Applied Science
10. Camacho EF, Bordons-Alba C (1995) Model predictive control in the process industry. Springer
11. Cellier FE, Kofman E (2006) Continuous system simulation. Springer
12. Chesmond CJ (1986) Control system technology. Edward Arnold
13. Considine DM (1985) Process instruments and controls handbook. McGraw-Hill
14. Crowder RM (1991) Electric drives and their controls. Clarendon Press
15. De Silva CW (2007) Sensors and actuators: control systems instrumentation. CRC Press
16. Dodds G (2005) Advanced process control unleashed. Instrumentation, Systems and Automation Society
17. Dorf RC, Bishop RH (2001) Modern control systems. Prentice-Hall
18. Engelbrecht AP (2007) Computational intelligence: an introduction. Wiley
19. Ferber J, Weiss G (1999) Multi-agent systems: an introduction to distributed artificial intelligence. Addison-Wesley
20. Fine TL (2006) Probability and probabilistic reasoning for electrical engineering. Prentice Hall
21. Fraden J (1997) Handbook of modern sensors: physics, designs, and applications. American Institute of Physics
22. Franklin GF, Powell JD, Emami-Naeini A (1994) Feedback control of dynamic systems. Addison-Wesley
23. Godfrey K (1981) Compartmental models and their applications. Academic Press

© ZRC SAZU/Research Centre of the Slovenian Academy of Sciences and Arts 2023
R. Karba et al., *Terminological Dictionary of Automatic Control, Systems and Robotics*,
Intelligent Systems, Control and Automation: Science and Engineering 104,
https://doi.org/10.1007/978-3-031-35755-8_27

24. Gasparyan ON (2008) Linear and nonlinear multivariable feedback control: a classical approach. Wiley
25. Hauptmann P (1993) Sensors: principles and applications. Prentice Hall
26. Hughes TA (1988) Measurement and control basics. Instrumentation, Systems and Automation Society
27. Isermann R, Lachmann K-H, Matko D (1992) Adaptive control systems. Prentice Hall
28. ISO/TR 20218-1 (2018) Robotics—safety design for industrial robot systems—part 1: end-effectors
29. ISO/TR 20218-2 (2018) Robotics—safety design for industrial robot systems—part 2: manual load/unload stations
30. ISO/TR 23482–2 (2019) Robotics—applications of ISO 13482—application guidelines
31. ISO TS 15066 (2016) Robots and robotic devices—collaborative robots
32. Johnson CD (1997) Process control instrumentation technology. Prentice-Hall International
33. Kailath T, Sayed AH, Hassibi B (2000) Linear estimation. Prentice Hall
34. Karer G, Škrjanc I (2013) Predictive approaches to control of complex systems. Springer
35. Khalil HK (2002) Nonlinear systems. Prentice Hall
36. Kissell TE (2000) Industrial electronics: applications for programmable controllers, instrumentation and process control, and electrical machines and motor controls. Prentice Hall
37. Klančar G, Zdešar A, Blažič S, Škrjanc I (2017) Wheeled mobile robotics, from fundamental towards autonomous systems. Butterworth-Heinemann
38. Kocijan J (2023) Modelling dynamic systems with artificial neural networks and related methods. University of Nova Gorica Press
39. Kulakowski BT, Gardner JF, Shearer JL (2007) Dynamic modelling and control of engineering systems. Cambridge University Press
40. Lenarčič J, Bajd T, Stanišić MM (2013) Robot mechanisms. Springer
41. Levine WS (ed) (2018) The control handbook. CRC Press
42. Ljung L (1999) System identification: theory for the user. Prentice Hall
43. Macfarlane AGJ (ed) (1980) Complex variable methods for linear multivariable feedback systems. Taylor & Francis
44. Maciejowski JM (1989) Multivariable feedback design. Addison-Wesley
45. Maciejowski JM (2002) Predictive control: with constraints. Prentice Hall
46. Matko D, Zupančič B, Karba R (1992) Simulation and modelling of continuous systems: a case study approach. Prentice Hall
47. Mihelj M, Bajd T, Ude A, Lenarčič J, Stanovnik A, Munih M, Rejc J, Šlajpah S (2019) Robotics. Springer
48. Mihelj M, Novak D, Beguš S (2014) Virtual reality technology and applications. Springer
49. Mihelj M, Podobnik J (2012) Haptics for virtual reality and teleoperation. Springer
50. Mikleš J, Fikar M (2007) Process modelling, identification, and control. Springer
51. Morris AS (1994) Principles of measurement and instrumentation. Prentice Hall
52. Murray-Smith DJ (2012) Modelling and simulation of integrated systems in engineering, issues, methodology, quality, testing and application. Woodhead Publishing Limited
53. Murray-Smith DJ (2015) Testing and validation of computer simulated models, principles, methods and applications. Springer
54. Mylroi MG, Calvert G (eds) (1984) Measurement and instrumentation for control. P. Peregrinus
55. Neelamkavil F (1989) Computer simulation and modelling. Wiley
56. Nelles O (2002) Nonlinear system identification. Springer
57. Ogata K (2010) Modern control engineering. Prentice-Hall
58. Patel RV, Munro N (1982) Multivariable system theory and design. Pergamon Press
59. Samad T, Annaswamy AM (2011) The impact of control technology. IEEE Control Systems Society
60. Sellers JJ, Astore WJ et al (2005) Understanding space: an introduction to astronautics. McGraw-Hill
61. Siciliano B, Khatib O, Kröger T (eds) (2008) Springer handbook of robotics. Springer
62. Siciliano B, Sciavicco L, Villani L, Oriolo G (2009) Modelling, planning and control. Springer

63. Skogestad S, Postlethwaite I (2005) Multivariable feedback control: analysis and design. Wiley
64. Slotine JJE, Li W (1991) Applied nonlinear control. Prentice Hall
65. Smith CA, Corpio AB (1997) Principles and practice of automatic process control. Wiley
66. Soloman S (1998) Sensors handbook. McGraw-Hill
67. Stenerson J (2005) Fundamentals of programmable logic controllers, sensors, and communications. Prentice-Hall
68. Strmčnik S, Juričić Đ (eds) (2013) Case studies in control: putting theory to work. Springer
69. Tan AH, Godfrey K (2019) Industrial process identification. Springer
70. Tangirala AK (2018) Principles of system identification: theory and practice. CRC Press
71. Urban G, Fraden J (2016) Handbook of modern sensors: physics, designs, and applications. Springer
72. Wang Q-G (2003) Decoupling Control. Springer
73. Wang Q-G, Ye Z, Cai W-J, Hang C-C (2008) PID control for multivariable processes. Springer
74. Wilson JS (2004) Sensor technology handbook. Elsevier

## *Terminological References*

75. Broadbent DT, Masubuchi M (eds) (2014) Multilingual glossary of automatic control technology: English-French-German-Russian-Italian-Spanish-Japanese. Elsevier
76. Cubberly WH (ed) (1988) Comprehensive dictionary of instrumentation and control: reference guides for instrumentation and control. Instrument Society of America
77. Dimon TG (2003) The automation, systems, and instrumentation dictionary
78. Institute of Electrical and Electronics Engineering (1989) IEEE standard glossary of modelling and simulation terminology
79. Jones CT, Jones PS (1996) Patrick-Turner's Industrial Automation Dictionary. Brilliant-Training
80. Rosenberg JM (1986) Dictionary of artificial intelligence and robotics. John Wiley & Sons, New York, Chichester etc.
81. Waldman H (1985) Dictionary of robotics. Collier Macmillan

# Correction to: Terminological Dictionary of Automatic Control, Systems and Robotics

**Correction to:**
**R. Karba et al.,** *Terminological Dictionary of Automatic Control, Systems and Robotics,* **Intelligent Systems, Control and Automation: Science and Engineering 104,**
**https://doi.org/10.1007/978-3-031-35755-8**

The original version of the book was inadvertently published with an incorrect abstract in the online version, which has now been corrected. The book has been updated with the change.

---

The updated version of the book can be found at
https://doi.org/10.1007/978-3-031-35755-8

© ZRC SAZU/Research Centre of the Slovenian Academy of Sciences and Arts 2024    C1
R. Karba et al., *Terminological Dictionary of Automatic Control, Systems and Robotics,*
Intelligent Systems, Control and Automation: Science and Engineering 104,
https://doi.org/10.1007/978-3-031-35755-8_28

Printed in the United States
by Baker & Taylor Publisher Services